D0928018

An Introduction to the Language of Mathematics

Frédéric Mynard

An Introduction to the Language of Mathematics

Frédéric Mynard
Mathematics
New Jersey City University
Jersey City, NJ, USA

ISBN 978-3-030-00640-2 ISBN 978-3-030-00641-9 (eBook)
https://doi.org/10.1007/978-3-030-00641-9

Library of Congress Control Number: 2018956569

Mathematics Subject Classification: 97E30, 97E40, 97E50, 97E60, 03B05, 03E20, 03E25, 03F07, 06A06, 11A05, 11A51

This Springer imprint is published by the registered company Springer Nature Switzerland AG
The registered company address is: Gewerbestrasse 11, 6330 Cham, Switzerland

To Dora, for all the love,
support, and inspiration.

Preface

This text is meant to be used as a textbook for a transition course from mathematics centered on calculation techniques to proof-based mathematics. Most US universities offer such a class after two semesters of Calculus, more for the sake of ensuring a minimum of mathematical maturity than because Calculus itself is of any help. Such a course often focuses on proof writing and is foundational for upper level mathematics courses. Proofs, however, are only a part of the *foundations* that need to be laid out to prepare for upper level courses. Hence, this text is about foundations and discusses not only proofs but the kind of system and conventions you need to build your proofs in. The radical change of perspective is often disorienting to students, and such courses often have high failure rates and leave students frustrated. For students who have come to think of mathematics as Calculus and its applications, it comes as somewhat of a shock that mathematics is about something else entirely. Here is what I want to tell them before engaging on that road:

This is a course unlike any other mathematics course you have taken so far. Where you have been focusing on basic techniques and calculations, you will focus on *arguments* from now on. Consider that you have been taking *pre-mathematics* courses until now and that you are taking a first *mathematics* (as what mathematicians do) course, where you will learn the language and way of thinking of mathematicians—mathematicians rather than mathematics—for mathematics is a human activity, driven by human impulses to understand and to model esthetically and efficiently.

This is a formal course, focused on formalism, though the goal of reaching fluency in reading proofs and proof making goes way beyond formalism. As such, the material is often disorienting for students at first and appears very dry and abstract. It is not unlike learning a foreign language from scratch: before you can see the beauty of poetry in that language, you have a long way to go in the sometimes tedious task of learning vocabulary and grammar, before developing the necessary intuition. Learning this material will probably be more difficult than any other mathematics class before, but it is worth the effort: you will sweat and puff pushing a very heavy door, but the land on the other side of the door is the secret garden of mathematicians, a land of beauty and harmony that you will surely enjoy exploring.

What matters at the end is mathematics in action: presenting beautiful arguments and proving meaningful statements, often surprising or far-reaching. But to get there, we have to start with basic vocabulary, *definitions*, and *concepts* that need to be assimilated like those of your mother tongue. That means you need to *know* definitions *precisely*, not approximately, not just by example. You should keep examples distinguishing various concepts in mind, but this is no substitute for a precise definition. A *proof* is an airtight argument within a certain logical system. Airtight is a pretty high standard when it comes to arguments. That means that you can only *prove* statements about things that are very specifically defined. You cannot *prove* anything at all about something whose *exact* meaning is unclear to you. I overemphasize this for a simple reason: many students will fail precisely because they remained content with "having an idea" of such and such a concept and did not make the effort to truly assimilate the *exact definition*. Without learning the basic vocabulary, you cannot even begin to learn a language or to pick up anything. This is no exception, so learning definitions carefully will be one of the keys to success in this course.

To learn a language well, it is not enough to learn its vocabulary, syntax, and grammar. You need to read a variety of styles and listen to native speakers in order to recognize standard patterns and gradually develop some intuition. Similarly, studying standard proofs carefully will be essential. It is not enough to understand each step of a proof; this would be enough to be convinced that the argument is valid and that this is indeed a proper proof, but it is inadequate with regard to the goal of training you to write your own proofs. To this end, you should keep going over a proof, thinking about how its parts are articulated, *until you are able to reproduce it on your own.*

Examples and exercises are drawn from a variety of sources, which are not specifically attributed but included in the references. The books [19] and [27] have been particularly influential sources. There are two kinds of exercises in this book. Those scattered through the text are an integral part of the course and should be attempted as you go. Full solutions for these exercises are provided at the end of the book. On the other hand, additional exercises without solutions are included at the end of most sections and can be used to assign homework. The instructor may request the solution manual for those exercises on the book's webpage

The reader will quickly notice that the text contains an unusual number of footnotes—a fact that may irritate some readers who would rather have everything incorporated in the text, and please others who will be happy to skip them altogether in a first reading, only to return to these additional comments in a second, more systematic, reading. The instructor may request the solution manual for those exercises on the book's webpage.

There are a few results that are stated or alluded to without proof in the text, because the arguments involved require more sophistication of the reader that one would expect of the target audience at this point in the course. These and related results are proved in Appendix A making the book self-contained.

To a large extent, Chapters 2, 3, and 4 are the core of the book, and Chapter 1 could be covered only lightly. It introduces the basics of logic and of the language of Set Theory, and as such, it is the natural place to start. Yet, in keeping with its opening remarks on language, it introduces an early, albeit superficial, discussion of the axioms of Set Theory—a choice that stems from my experience that semi-philosophical considerations in class discussions often turn out to be very fruitful, but a choice that some instructors may prefer not to follow. It should not be hard for such instructors to recast Section 1.4 under a more naive light to fit their needs. Section 1.7 is the other part of Chapter 1 that one might find a little ambitious when compared to the opening of Chapter 2. The choice to introduce the notions of one-to-one, onto, and bijective maps early on (in that section) in the context of finite sets to illustrate some counting arguments from the set-theoretic viewpoint has also proved useful in my experience, when it is time to return to these notions in the context of infinite cardinalities (Chapter 4). Again, it would not be difficult for an instructor to choose to treat this material (Section 1.7) later, for instance, when treating functional relations in Section 3.1.1. I should note that Chapter 3 treats relations a little more extensively than other similar books and includes a number of informal comments on their nature, particularly regarding equivalence relations. I have found such informal discussions to be an important complement to the formal development of the material, but here again, the instructor can easily skip what she sees as too chatty.

I am grateful to the readers of early drafts who caught a number of typos and errors, particularly my students at NJCU, among which Fadoua Chigr and John Stulich stand out for the meticulosity of their reading. My colleagues Deborah Bennett (NJCU) and Szymon Dolecki (University of Burgundy) made many helpful suggestions, and so did the anonymous referees for Springer. I am indebted to all of them. Most likely, many imperfections remain, and they are of course my sole responsibility. Finally, I am also grateful to New Jersey City University, for extensive parts of this manuscript were written while I was benefiting from a course release under the Separately Budgeted Research program.

Jersey City, NJ, USA Frédéric Mynard
June 29, 2018

Contents

Chapter 1
The Language of Logic and Set-Theory

1.1 A Language for Proofs?

We set out to develop a language in which we can write *proofs*, that is, airtight arguments showing the validity of a statement. In order to do that, all terms of the statement (and of the proof) need of course to be unequivocally defined, which is not as easy to achieve as you may think at first. Let us start with reflecting upon what it would take, as a way of motivating a treatment of the material that you will likely find very formal and abstract.

If we want all terms involved in a mathematical statement and its proof to be well-defined, we are immediately faced with the problem that common language does not provide *any* unambiguous definition, that is, all definitions are context dependent and subject to interpretation. Indeed, if you look up a word in the dictionary, it is defined in terms of other words, which you may in turn look up in your dictionary, and this process will eventually turn circular, getting you back where you started. You will never get to foundational terms that are not defined in terms of other words.

That is to say, the meaning of a given word cannot be understood independently of the web of interrelations between words, which is what in effect "defines" each term, but there is no *primitive word* forming a foundation of language, in terms of which all other words would be defined. Hence, in common language *meaning* depends on ever changing relations among words and is heavily context dependent. The *linguistic turn* in philosophy during the 20th century has much to do with exploring the consequences of this simple observation.

A precise static meaning would require a fixed collection of primitive foundational words that are not defined in terms of other words but "stand on their own." There are no such notions in common language.

F. Mynard, *An Introduction to the Language of Mathematics*,
https://doi.org/10.1007/978-3-030-00641-9_1

As a consequence, common language cannot be enough for writing proofs and thus we need to develop a *deductive theory,* that is, an alternative adequate language to formulate our arguments. Any such theory is necessarily founded on *primitive* notions, that is, undefined notions—for otherwise, a notion could be defined modulo an infinite string of other notions without ever being able to rest on a concept whose meaning stands on its own.

To give *meaning* to primitive notions, unproven propositions called *axioms* declare true certain facts regarding the primitive concepts. The meaning of a primitive notion is thus delineated by the axioms that involve it, that is, the properties we ask of the object are what defines it. Nonprimitive notions of the theory are defined from primitive notions modulo specific syntactic rules. Propositions of the theory are obtained from axioms modulo logical inferential rules.

If all this sounds very formal, it's because it is. But the point of this discussion is that there is no way around some level of formality if we are to develop an adequate language for proofs. Thus, this first chapter will be largely devoted to the syntax and rules of logic, and their interpretation in the context of Set Theory.

In Set Theory, which constitutes the foundation of all mathematics, the notion of *set* is primitive, and so is the membership relation $\in$, where the expression

$$x \in A$$

reads as "x is a *member* of the set A" or "x is an *element* of the set A." We may alternatively say that "*x belongs to A*." We write

$$x \notin A$$

for the negation of $x \in A$. Hence, for instance, if $\mathbb{N}$, $\mathbb{Z}$, $\mathbb{Q}$, and $\mathbb{R}$ denote, respectively, the set of natural numbers, integers, rationals, and real numbers, respectively, then

$$1 \in \mathbb{N} \text{ but } -1 \notin \mathbb{N},$$

that is, 1 is a member of the set $\mathbb{N}$ but -1 is not. Similarly, $-1 \notin \mathbb{N}$ but $-1 \in \mathbb{Z}$ and $\frac{1}{2} \notin \mathbb{Z}$ but $\frac{1}{2} \in \mathbb{Q}$.

As already noted, the actual meaning of the word *set* and of the expression "$x \in A$" is determined by the axioms of Set Theory, some of which we will examine in Section 1.4, after we introduce the language of logic necessary to formulate them.

For the time being, we will simply rely on the intuition (as opposed to a definition) that a set is a collection of elements, so that *a set is determined by its elements*—a fact formalized by the axiom of extensionality (see page 21). Hence a set might be given by the list of its elements, given between *set builder braces* { and }. For instance,

$$\{1,2,a,\triangle\}$$

denotes the set whose elements are 1, 2, a, and $\triangle$, so that

$$\triangle \in \{1,2,a,\triangle\} \text{ but } 0 \notin \{1,2,a,\triangle\}.$$

Since elements determine the set, giving the list of elements between set builder parenthesis in a different order does not affect the set. For instance,

$$\{1,2,a,\triangle\} = \{\triangle,a,2,1\}.$$

More generally $A = B$ means that the set A and B have exactly the same elements.

Exercise 1.1. How many elements do the following sets have:

1. $\{a,b,c,b,a\}$;
2. $\{a,b,\{c,d\}\}$;
3. $\{a,b,\{a,b\}\}$.

Set builder notation can also be used to consider the set of elements of a given set that satisfy a given property. For instance,

$$\{x \in \mathbb{R} : x^2 = 4\}$$

denotes the set of real numbers x such that $x^2 = 4$, (that such expressions always define sets is also an axiom, called axiom of separation, introduced on page 21 of real numbers x such that $x^2 = 4$), in other words,

$$\{x \in \mathbb{R} : x^2 = 4\} = \{-2,2\}.$$

Note that the colon in this convention reads as "such that."

As we have seen, if x is an element of the set A, we write $x \in A$, in which one should keep in mind that the relation $\in$ is also primitive.

Starting from the elementary formulas $x \in X$ and $X = Y$ we build further formulas from logical connectives: the negation $\neg$, disjunction $\vee$, conjunction $\wedge$, implication $\Longrightarrow$, equivalence $\Longleftrightarrow$, and the existential quantifier $\exists$, which reads as "there exists," and the universal quantifier $\forall$, which reads as "for each." We will clarify the meaning of each as we move along.

Additional Exercises (Homework)

Exercise 1.2. How would you write the set whose elements are a, b, 1, the set whose elements are a and 2, and the set whose only element is the set whose only element is 4?

Exercise 1.3. How many elements does the set

$$\{\{1\},\{\{\{1\}\}\}\}$$

have?

1.2 Propositional Logic

1.2.1 Propositions

Logical connectives apply to *logical propositions.*

Definition 1.4. A *proposition* is a statement with a well-defined truth value, either true (T) or false (F).

A proposition cannot be both or neither.

Example 1.5. A statement such as

$$4^2 = 16$$

is a proposition whose truth value is T while

$$1 = 0$$

is a proposition whose truth value is F. On the other hand, the expression

$$x^2 = 16$$

is **not** a proposition as stated, for the truth value of the formula is undetermined until the value of x is. Similarly, "she is taller than 5ft 8 inches" is **not** a proposition, unless who "she" refers to is clearly specified. Neither is the statement

$$\text{this sentence is false.} \tag{1.1}$$

Indeed, this statement cannot be assigned a truth value, for if (1.1) is true, then it is false, and if it is false, then it is true. Such statements are *paradoxes* and are excluded from our context of propositional logic.

A statement such as

life will be extinct on Earth by 3000

is a proposition (it has a unique truth value), even though its truth value will not be known for a long time, potentially until the year 3000.

We will often consider propositional variables p, q, r, etc., that is, propositions that are not specified, beyond the fact that they are propositions, hence are either true or false (and not both). We use them as arguments (that is, as variables) of logical connectives to form new *propositional forms*, that is, formulas that include propositional variables and become propositions if the truth values of each propositional variable is assigned.

To clarify the meaning of the logical connectives, we give explicitly their truth tables in terms of the truth values (T for True, F for False) of their arguments (which are propositional variables).

1.2.2 Basic Logical Connectives

Definition 1.6. The *negation* $\neg$ applies to a single proposition or propositional variable:

p	$\neg p$
T	F
F	T

The table above, called *truth table*, indicates the truth value of the propositional form $\neg p$ as a function of the truth value of the propositional variable p. When considering a propositional form of more than one variable, the truth table needs to show the assignment of truth value to the propositional form for every possible combination of truth values of the arguments.

Definition 1.7. *Disjunction* $\vee$ connects two propositions or propositional variables (in a symmetric way) and reads as "or," where "or" is inclusive, that is, $p \vee q$ means p or q or both, that is, $p \vee q$ is true if *at least one* of p and q is true. Similarly, *conjunction* $\wedge$ connects two propositions or propositional variables (in a symmetric way) and reads as "and," so that $p \wedge q$ is true only if both p and q are. In terms of truth tables, we have ([1])

p	q	$p \vee q$	$p \wedge q$
T	T	T	T
T	F	T	F
F	T	T	F
F	F	F	F

Example 1.8. The negation of the statement "3 divides 9" (which is true) is "3 does not divide 9" (which is false). The negation of "All humans are mammals" (which is true) is "it is not the case that all humans are mammals," equivalently, "there is a human that is not a mammal" (which is false).

Example 1.9. If, for instance, p is "15 is prime" (which is false) and q is "36 is a multiple of 3" (which is true), then $p \vee q$ reads as "15 is prime or 36 is a multiple of 3" and is true, for at least one of the arguments (here, q) is true. On the other hand, $p \wedge q$ reads as "15 is prime and 36 is a multiple of 3" and is false, for at least one of the arguments (here, p) is false. If r is "15 is a multiple of 5" (which is true), then $q \vee r$, which reads "36 is a multiple of 3 or 15 is a multiple of 5," is true for disjunction is an *inclusive* form of "or."

[1] Note that the truth table accounts for all four possible combinations of truth values for the pair (p,q).

Exercise 1.10. Formulate the following statements in terms of the propositions of Example 1.9 and the connectives $\vee, \wedge$, and $\neg$, and give their truth value.

1. "15 is not prime but 36 is a multiple of 3";
2. "Although 15 is a multiple of 5, 36 is not a multiple of 3";
3. "While 36 is a multiple of 3, 15 is not prime"

Remark 1.11. As seen in this exercise, "but," "although," and "while" often logically translate to "and."

1.2.3 First Laws of Propositional Logic

Two propositions are *equivalent* if they have the same truth value, so that "$2 = 1$" and "15 is prime" are equivalent propositions, because they are both false.

We say that two propositional forms are *equivalent* if they are identical as functions of the truth values of their arguments, that is, if their respective columns in a truth table are identical. When two propositional forms P and Q are equivalent, we write

$$P \equiv Q.$$

Definition 1.12. A propositional form is a *tautology* if it is True for every assignment of truth values of its arguments, and a *contradiction* if it is False for every assignment of truth values of its arguments.

We write $P \equiv \mathrm{T}$ if P is a tautology and $P \equiv \mathrm{F}$ if P is a contradiction. In other words, in adopting this notation, we consider the constant functions T and F (of any set of propositional variables) as propositional forms. A *law* of propositional logic is an equivalence of propositional forms. A law is thus true in form, regardless of the truth values of its arguments, that is, every instantiation of a law is true.

Example 1.13. Of course, $p \vee \neg p$ is a tautology. This fact is often called *Law of excluded middle*, because it states that a proposition or its negation has to be true. There is no third possibility. Similarly, $p \wedge \neg p$ is a contradiction, for p and $\neg p$ cannot both be true simultaneously.

p	$\neg p$	$p \vee \neg p$	$p \wedge \neg p$
T	F	T	F
F	T	T	F

Exercise 1.14. Show that $(p \vee q) \vee (\neg p \wedge \neg q)$ is a tautology and $(p \vee q) \wedge (\neg p \wedge \neg q)$ is a contradiction.

There are equivalences of propositional forms that are immediate from the definitions. Namely,

$$p \vee q \equiv q \vee p \text{ and } p \wedge q \equiv q \wedge p, \qquad \text{(commutative laws)}$$

and

$$\neg(\neg p) \equiv p. \qquad \text{(double negation law)}$$

Only slightly less immediate is the fact that $\vee$ and $\wedge$ are associative, which is easily verified either by using a truth table or by noting that $p \wedge (q \wedge r)$ is true only if p and $(q \wedge r)$ are true, and that the latter is true only if q and r are true. Thus $p \wedge (q \wedge r)$ is true only if all three of p, q, and r are true, and similarly for $(p \wedge q) \wedge r$. Analogously, both $p \vee (q \vee r)$ and $(p \vee q) \vee r$ are true if and only if at least one of p, q, and r is true. In other words,

$$p \wedge (q \wedge r) \equiv (p \wedge q) \wedge r \text{ and } p \vee (q \vee r) \equiv (p \vee q) \vee r. \qquad \text{(associative laws)}$$

The following laws provide useful equivalent forms for the propositional forms of the negation of conjunction and disjunction.

Proposition 1.15 (De Morgan's Laws).

$$\neg(p \vee q) \equiv \neg p \wedge \neg q$$
$$\neg(p \wedge q) \equiv \neg p \vee \neg q.$$

Proof. This follows from examining the following truth table:

p	q	$p \vee q$	$p \wedge q$	$\neg(p \vee q)$	$\neg p \wedge \neg q$	$\neg(p \wedge q)$	$\neg p \vee \neg q$
T	T	T	T	F	F	F	F
T	F	T	F	F	F	T	T
F	T	T	F	F	F	T	T
F	F	F	F	T	T	T	T

□

Of course, while a truth table makes the fact plain, you may want to convince yourself by other means. For instance, note that $p \vee q$ is false if and only if p and q are both false, that is, if $\neg p$ and $\neg q$ are both true. In other words, $\neg(p \vee q)$ is true exactly if $\neg p$ and $\neg q$ is true, that is, exactly if $\neg p \wedge \neg q$ is.

Remark 1.16. Note that in particular, this shows that the forms in Exercise 1.14 are in fact of the form $p \vee \neg p$ and $p \wedge \neg p$, respectively, and are thus clearly a tautology and a contradiction, respectively.

Example 1.17. Given a specific function $f : \mathbb{R} \to \mathbb{R}$, consider the statement $r =$"f is decreasing and concave down." It has the form $p \wedge q$ where p is "f is decreasing" and q is "f is concave down." Suppose that r is false, that is, $\neg r$ is true. Moreover, $\neg r$ reads as "it is not the case that f is decreasing and concave down." We can reformulate this in a more straightforward way via De Morgan's law: $\neg(p \wedge q) \equiv \neg p \vee \neg q$, that is, $\neg r$ is "f is not decreasing or f is not concave down."

A *denial of a proposition* p is a proposition equivalent to $\neg p$, that is, a proposition with the same truth value as $\neg p$. Similarly, a denial of a propositional form P is a propositional form equivalent to $\neg P$.

Example 1.18. A denial of "Either Colonel Mustard is innocent or the crime did not take place in the patio" can be obtained using De Morgan's laws. Let c denote "Colonel Mustard is guilty" or if you prefer "Colonel Mustard is not innocent," and let p denote "the crime took place in the patio." The statement to deny is then of the form $\neg c \vee \neg p \equiv \neg(c \wedge p)$ so that, by the law of double negation, a denial is of the form $c \wedge p$, that is, "Colonel Mustard is guilty and the crime took place in the patio."

Proposition 1.19 (Distributive Laws).

$$p \wedge (q \vee r) \equiv (p \wedge q) \vee (p \wedge r) \tag{1.2}$$

$$p \vee (q \wedge r) \equiv (p \vee q) \wedge (p \vee r). \tag{1.3}$$

Proof. This can be verified straightforwardly by examination of the truth tables. We only check the first distributive law and the other is verified similarly—which you should do as an exercise.

p	q	r	$p \wedge r$	$p \wedge q$	$(p \wedge q) \vee (p \wedge r)$	$q \vee r$	$p \wedge (q \vee r)$
T	T	T	T	T	T	T	T
T	T	F	F	T	T	T	T
T	F	T	T	F	T	T	T
T	F	F	F	F	F	F	F
F	T	T	F	F	F	T	F
F	T	F	F	F	F	T	F
F	F	T	F	F	F	T	F
F	F	F	F	F	F	F	F

□

Remark 1.20. Note that in the previous example, some propositional forms (such as $p \wedge (q \vee r)$) depend on 3 arguments, each of which can take two values T or F. Hence, the truth table needs to account for all of the 2^3 possibilities for the truth values of (p, q, r), resulting in 8 rows. More generally, the truth table for a propositional form that depends on n propositional variables will have 2^n rows.

Remark 1.21. The distributive laws are formally similar to the distributive law of algebra:

$$a \cdot (b + c) = a \cdot b + a \cdot c$$

is formally identical with the first distributive law if we denote $\wedge$ by $\cdot$ and $\vee$ by $+$, which some logicians do. Note however that the analogy is only partial: there is no algebraic analog to the logical distributive law (1.3) if we replace $\wedge$ by $\cdot$ and $\vee$ by $+$, because there are numbers a, b, and c such that

$$a + (b \cdot c) \neq (a + b) \cdot (a + c).$$

Exercise 1.22. Consider the propositions of Example 1.18, and let w denote "the crime weapon was the knife." How would $p \wedge (c \vee w)$ read? How would its negation read?

Exercise 1.23. Build truth tables for the following propositional forms:

1. $p \vee (q \wedge \neg r)$
2. $(\neg q \vee p) \wedge (r \vee \neg s)$

1.2.4 Conditional and Biconditional Statements

Definition 1.24. Implication $\Longrightarrow$ connects two propositions or propositional variables in an asymmetric way: in $p \Longrightarrow q$, the proposition p is the *premise*, or *antecedent*, and q is the *conclusion* or *consequent*. Saying that $p \Longrightarrow q$, which reads as "p implies q" or "if p then q," means that if p is true, q is necessarily true. In other words, $p \Longrightarrow q$ is false only if the premise is true but the conclusion is not:

p	q	$p \Longrightarrow q$
T	T	T
T	F	F
F	T	T
F	F	T

You are familiar with conditional statements such as "If f is differentiable at x, then f is continuous at x" (where this statement should in fact be preceded by "for every $f : \mathbb{R} \to \mathbb{R}$ and every $x \in \mathbb{R}$" to make it a proposition), in which the two propositions have related *meaning*. You are also used to consider such a statement only in the case where the premise is true to conclude that the conclusion is also true, but you normally do not pay attention to the truth value of the conditional statement for other combinations of truth values of the premise and conclusion. In other words, you usually use $p \Longrightarrow q$ in the context of what is sometimes called *modus ponens*: If both p and $p \Longrightarrow q$ are true, then we can conclude that q is true. This is usually (but not always) how conditional statements are also used in proofs. In particular you will often see expressions that take one of the forms "q because p," "Since p, q," "q, for p," "p, so q" or their variants. Note that in each case, two things are stated: that p is known to be true, and that q is a consequence, that is, each is a shorthand for modus ponens: p and p implies q, therefore q. Yet it is important to also have in mind what $p \Longrightarrow q$ means outside of this context.

While this is the very definition, let me emphasize again that $p \Longrightarrow q$ is false *only if* p is true and q is false. Consider, for instance, a political promise such as "if the deficit reaches 500% of GDP, I will raise taxes on the rich." The politician uttering such a statement did not lie if the deficit does not reach 500% of GDP, regardless of what she does about taxes. That means that we might easily consider true statements of the form $p \Longrightarrow q$ in which p and q have unrelated meanings. For instance,

If it snows at the surface of the sun then the USA
will win all forthcoming soccer world cups

or

if 4 is odd then 8 is prime

are true statements, because they are conditional statements in which the premise is false. Regardless of the truth value of the conclusion, the conditional statement is true. Hence $p \Longrightarrow q$ is true whenever p is false. Similarly, $p \Longrightarrow q$ is true whenever q is true, so

if π is an integer then Paris is the capital of France

is a true statement, because the conclusion of the conditional statement is true.

Thus we have seen that $p \Longrightarrow q$ is true whenever p is false or q is true. In other words $p \Longrightarrow q$ is logically equivalent to $q \vee \neg p$:

$$(p \Longrightarrow q) \equiv (q \vee \neg p), \tag{1.4}$$

as explicitly shown by the truth table below, in which we also examine $\neg q \Longrightarrow \neg p$:

p	$\neg p$	q	$\neg q$	$\neg p \vee q$	$p \Longrightarrow q$	$\neg q \Longrightarrow \neg p$
T	F	T	F	T	T	T
T	F	F	T	F	F	F
F	T	T	F	T	T	T
F	T	F	T	T	T	T

Hence the negation of a conditional statement can be understood modulo (1.4) and Proposition 1.15:

$$\begin{aligned} \neg(p \Longrightarrow q) &\equiv \neg(\neg p \vee q) \text{ by (1.4)} \\ &\equiv \neg\neg p \wedge \neg q \text{ by Proposition 1.15} \\ &\equiv p \wedge \neg q, \end{aligned}$$

so that

$$\neg(p \Longrightarrow q) \equiv p \wedge \neg q. \tag{1.5}$$

In other words, to prove $\neg(p \Longrightarrow q)$, that is, to prove $p \Longrightarrow q$ false, we need to find the premise p true *and* the conclusion q false, as already noted from the very definition of a conditional statement.

Note that the truth table above proves

$$(p \Longrightarrow q) \equiv q \vee \neg p \equiv (\neg q \Longrightarrow \neg p).$$

Definition 1.25. The last form $\neg q \Longrightarrow \neg p$ is called the *contrapositive* of $p \Longrightarrow q$ and is logically equivalent to $p \Longrightarrow q$:

$$(p \Longrightarrow q) \equiv (\neg q \Longrightarrow \neg p). \tag{1.6}$$

The *converse* of $p \Longrightarrow q$ is $q \Longrightarrow p$.

Note that the converse of a conditional statement is **not** equivalent to that statement, as illustrated in the following:

Exercise 1.26. Consider the true statement from Calculus "If a function f is differentiable at a real number x, then f is continuous at x."

1. Write it explicitly as a conditional statement;
2. Write its converse. Is the converse true?
3. Write its contrapositive. Is the contrapositive true?
4. Write the negation of this statement.

Remark 1.27. Of course, $p \Longrightarrow q$ and $q \Longrightarrow p$ may have the same truth value *for some instances of p and q*, but one cannot be inferred from the other in general.

Exercise 1.28. State the converse and contrapositive of the conditional statement "If it snows tonight, then I will stay home tomorrow."

Exercise 1.29. Show that

$$((p \Longrightarrow q) \wedge (q \Longrightarrow r)) \Longrightarrow (p \Longrightarrow r)$$

is a tautology, that is, is always true.

Of course, it is expected that if p implies q and q implies r then p implies r, that is, $\Longrightarrow$ is transitive.

Recall that:

Definition 1.30. Propositions p and q are *equivalent*, denoted $p \iff q$, if they have the same truth values.

The relation between $p \iff q$, $p \Longrightarrow q$, and $q \Longrightarrow p$ are summarized in the truth table below:

p	q	$p \Longrightarrow q$	$q \Longrightarrow p$	$(p \Longrightarrow q) \wedge (q \Longrightarrow p)$	$p \iff q$
T	T	T	T	T	T
T	F	F	T	F	F
F	T	T	F	F	F
F	F	T	T	T	T

Note that

$$(p \iff q) \equiv (p \Longrightarrow q) \wedge (q \Longrightarrow p), \tag{1.7}$$

which is often how we prove equivalences, by proving two conditional statements.

Exercise 1.31. Show that propositional forms P and Q are equivalent ($P \equiv Q$) precisely when $P \iff Q$ is a tautology.

Remark 1.32. Note that there are various ways to say $p \Longrightarrow q$ or $p \Longleftrightarrow q$ in plain English. For instance, "if p then q," "p implies q," "p is sufficient for q," "q whenever p," "q when p," "p only if q," and "q is necessary for p" are all ways to state

$$p \Longrightarrow q.$$

Similarly,

$$p \Longleftrightarrow q$$

can be stated by "p if and only q" (p iff q), "p is equivalent to q," "p is necessary and sufficient for q," or "q is necessary and sufficient for p."

Exercise 1.33. Restate as a conditional propositional form ([2]) "X is compact is sufficient for X to be bounded" and "A necessary condition for a group G to be cyclic is that G be Abelian."

Exercise 1.34. Consider the statement "If p is a prime number that divides $a \cdot b$, then p divides a or b." Introduce appropriate propositions to give a propositional form for this statement. ([3])

Additional Exercises (Homework)

Exercise 1.35. Which of the following are propositions? Give the truth value of each proposition.

1. What time is it?
2. It is not the case that e^{π} is not rational.
3. $5x - 2y$ is an integer.
4. Either π is rational or e is rational.
5. Either π and e are irrational, or 2 is irrational.
6. The truth value of this statement is false.

Exercise 1.36. Make truth tables for

1.

$$(p \wedge q) \vee (r \wedge \neg s)$$

2.

$$((p \Longrightarrow q) \wedge p) \Longrightarrow q$$

3.

$$(p \Longrightarrow q) \vee (q \Longrightarrow p)$$

---

[2]In this exercise, we focus on the form of statements, and it is not needed to know what "compact," "bounded," "group," "cyclic" or "Abelian" actually mean.

[3]We are only concerned with form here. However, this is a true statement known as Euclid's Lemma. See Corollary 82 for a proof.

Exercise 1.37. Suppose that P is a tautology and Q is a contradiction. Are the following tautologies, contradictions, or neither?

1. $P \vee Q$;
2. $P \wedge Q$;
3. $\neg(P \vee Q)$;
4. $\neg(\neg P \wedge Q)$.

Exercise 1.38. How would a denial of

"balloons are red and ribbons are blue"

read?

Exercise 1.39. Given a specific function f, how would a denial of

"f has a local maximum at $x = 2$ or a local minimum at $x = 4$"

read?

Exercise 1.40. Write a truth table for XOR, the exclusive or of "soup or salad." Then show that

$$p\text{XOR}q \equiv (p \vee q) \wedge \neg(p \wedge q).$$

Exercise 1.41. Introduce appropriate propositions to give the logical form of the following statements:

1. Cristiano Ronaldo is a Portuguese player, a superstar, but a bad loser.
2. Bob and Laura are persuasive, though neither is honest.
3. I will pick up the phone if and only if I'm home but not asleep.

Exercise 1.42. Introduce appropriate propositions to give the logical form of the statement "Not only will John and Mary have to live on campus if they are not local residents, but neither will be allowed to park on campus—whether or not they are local residents."

Exercise 1.43. Introduce appropriate propositions to give the logical form of the statement "The players will go back to work if agreement is reached about their salaries, but this will be achieved, if at all, only if some of them take early retirement."

Exercise 1.44. Introduce appropriate proposition to give the propositional form of the following statements as a conditional or biconditional statement:

1. A time of less than 3 minutes is necessary to qualify for the Olympics;
2. A time of less than 3 minutes is sufficient to qualify for the Olympics;
3. The fish bite only at night;
4. A function is bounded whenever it is integrable;
5. Differentiability is sufficient for continuity;

6. a divides 5 only if a divides 25;
7. For two triangles to be congruent it is necessary and sufficient that they have 3 sides in common;
8. I will go for a run unless it rains.

1.3 Quantifiers and Quantified Statements

1.3.1 Predicates, Universe of Discourse, and Truth Sets

As we have seen, an expression such as

$$x^2 - 16 = 0 \tag{1.8}$$

is not a proposition, for x is unspecified. This is what we call a predicate:

Definition 1.45. A *predicate* is an expression that contains one or more variables, that becomes a proposition when the variables are replaced by specific objects.

For instance, (1.8) is a predicate $P(x)$ of the variable x. $P(x)$ is not a proposition, but $P(1)$ is a (false) proposition, and $P(4)$ is a (true) proposition.

Definition 1.46. The *truth set* of a predicate is the set of values assumed by the variable or variables for which the predicate becomes a true proposition.

Of course, to talk about the truth set of a predicate, we first need to fix a *universe of discourse*, that is, the set in which the variables are taking values ([4]).

Example 1.47. For instance, if we fix $\mathbb{Z}$ as the universe of discourse (the values among which x ranges), then the truth set of (1.8) is $\{-4,4\}$, while if we fix $\mathbb{N}$ to be the universe of discourse, then the truth set of (1.8) is $\{4\}$. The truth set of the predicate

$$\text{“}x^2 = 2\text{”}$$

is empty if the universe of discourse is $\mathbb{Q}$, and is $\{-\sqrt{2}, \sqrt{2}\}$ if the universe of discourse is $\mathbb{R}$.

Definition 1.48. For a given universe of discourse, two predicates $P(x)$ and $Q(x)$ are *equivalent*, in symbols,

$$P(x) \equiv Q(x),$$

if they have the same truth set.

[4]Note that we are here adopting terminology consistent with [27] or [10], while some logicians would reserve the term “universe” for something else, and use the term *domain* for “universe of discourse” and *domain of definition* for “truth set.”

For instance, if the universe of discourse is $\mathbb{N}$, then

$$\left(x^2 - 16 = 0\right) \equiv (x = 4),$$

but these two predicates are *not* equivalent if the universe of discourse is $\mathbb{R}$.

Remark 1.49. Let me remind you of a familiar procedure to "lift" an operation from a set of numbers to a set of functions. Consider, for instance, real-valued functions $f : \mathbb{R} \to \mathbb{R}$ and $g : \mathbb{R} \to \mathbb{R}$. The sum of the functions is defined pointwise by

$$(f+g)(x) = f(x) + g(x),$$

where the addition on the right-hand side is performed in the reals, because once the functions are evaluated, the values $f(x)$ and $g(x)$ are in $\mathbb{R}$. Since predicates are functions taking values in a set of propositions, we can apply logical connectives to predicates in a similar fashion: if $P(x)$ and $Q(y)$ are predicates, then $P(x) \wedge Q(y)$ denotes the predicate of the variables x and y assigning to (x, y) the proposition $P(x) \wedge Q(y)$, where $\wedge$ is here applied to the propositions $P(x)$ and $Q(y)$ obtained for specific instances of x and y. It is in this sense that logical connectives are applied to predicates in what follows.

1.3.2 Existential and Universal Quantifiers

In many situations, what matters about the truth set is whether it is non-empty (*existence* of solution), or if it is the entire universe of discourse (generally true statement, or *universal* statement).

Definition 1.50 (Existential Quantifier). For a given universe of discourse and a given predicate $P(x)$, the expression

$$\exists x\, P(x)$$

reads as "there exists x such that $P(x)$" or "for some x, $P(x)$" and is a proposition that is true if and only if the truth set of $P(x)$ is non-empty.

Definition 1.51 (Universal Quantifier). For a given universe of discourse and a given predicate $P(x)$, the expression

$$\forall x\, P(x)$$

reads as "for all x, $P(x)$" and is a proposition that is true if and only if the truth set of $P(x)$ is the entire universe of discourse.

Remark. Note that the symbol $\exists$ looks like the mirror image of E, like in "Exists." Similarly, $\forall$ looks like an upside down A, like in "All."

Exercise 1.52. If the universe of discourse is $\mathbb{R}$ what are the truth values of the following propositions:

1. $\exists x\ (x \leq 3)$.
2. $\forall x\ (x+1 > x)$.
3. $\forall x\ (|x| > 0)$.
4. $\exists x\ (x < 3 \wedge 2x - 4 = 0)$. Recall here Remark 15, and note that we assert here the existence of the same instance of x satisfying $x < 3$ and $2x - 4 = 0$.
5. $\exists x\ \left(x^2 + 9 = 0\right)$.
6. $\forall x\ (x < 3 \wedge 2x - 4 = 0)$.

Example 1.53. Suppose we want to formalize a statement such as

$$\text{“all trees produce oxygen.”} \tag{1.9}$$

Let $O(x)$ denote the predicate "x produces oxygen." If the universe of discourse is the set of trees, then our statement (1.9) takes the form

$$\forall x\, O(x).$$

On the other hand, if the universe of discourse is the set of plants, then we need to introduce a new predicate $T(x)$ denoting "x is a tree" and (1.9) takes the form

$$\forall x\, (T(x) \Longrightarrow O(x)),$$

that is, "for all plant x, if x is a tree, then x produces oxygen."

You might be tempted to use

$$\forall x\, (T(x) \wedge O(x))$$

for (1.9) instead in this context, but note that this is a completely different statement, namely that all plants are trees and produce oxygen!

Similarly, if we want to formalize

$$\text{“some trees produce oxygen”} \tag{1.10}$$

the propositional form used depends on the universe of discourse. If the universe of discourse is the set of trees, we have

$$\exists x\, O(x).$$

If, on the other hand, the universe of discourse is the set of plants, then (1.10) takes the form

$$\exists x\ (T(x) \wedge O(x)),$$

that is, "there is a plant that is a tree and produces oxygen."

More generally, note that a statement of the form "All $P(x)$ satisfy $Q(x)$" takes the form

$$\forall x\, (P(x) \Longrightarrow Q(x)),$$

while a statement of the form “Some $P(x)$ satisfy $Q(x)$” takes the form

$$\exists x\,(P(x)\wedge Q(x))\,.$$

Definition 1.54 (Restriction to a Subset of the Universe of Discourse). If $P(x)$ is a predicate, X is the universe of discourse, and S is a subset ([5]) of X (or equivalently the truth set of a second predicate $S(x)$), then we abbreviate the forms $\forall x\,(S(x)\Longrightarrow P(x))$ and $\forall x\,(x\in S\Longrightarrow P(x))$ (stating that every x in S satisfy $P(x)$) as

$$\forall x\in S\;P(x).$$

Similarly, we abbreviate the forms $\exists x\,(S(x)\wedge P(x))$ and $\exists x\,(x\in S\wedge P(x))$ (stating that some x in S satisfies $P(x)$) as

$$\exists x\in S\;P(x).$$

Example 1.55. In a universe of discourse formed by all complex numbers, a statement such as “for every natural number, there is a larger real number” should normally be formalized as

$$\forall n\,(n\in\mathbb{N}\Longrightarrow\exists r\,(r\in\mathbb{R}\wedge r>n))\,,$$

but, with the convention of Definition 1.54, we can simplify this to

$$\forall n\in\mathbb{N}\,\exists r\in\mathbb{R}\;(r>n)\,.$$

Sometimes you will find the notation

$$\exists!\,P(x)$$

which reads as “there is *a unique* x (in the universe of discourse) satisfying $P(x)$.” Formally,

$$\exists! x\,P(x)=(\exists x\,P(x))\wedge(\forall y,z\,(P(y)\wedge P(z))\Longrightarrow y=z)\,,\tag{1.11}$$

that is, existence of a unique element of the truth set is two fold: existence ($\exists x\,P(x)$) *and* uniqueness ($\forall y,z\,(P(y)\wedge P(z))\Longrightarrow y=z$). Note that the latter means that for every possible pair (y,z) of elements of the universe of discourse, if $P(y)$ and $P(z)$ are both true, then $y=z$, that is, if P is true for two values of the variable, these values are actually the same. In other words, P is true for at most one value of the variable.

[5] We anticipate the definition of a subset in Section 1.5: S is a subset of X if every element of S is an element of X, that is,

$$\forall x\,(x\in S\Longrightarrow x\in X)\,.$$

1.3.3 A Word on Syntax

Even if some sources, such as [25], sometimes place quantifiers at the end of a formula, this is improper!

Quantifiers ($\exists$ and $\forall$) **come first** and modify **only** what follows them in an expression, and the assertion comes at the end, after all quantifiers. This statement needs to be qualified: the *scope* of the quantifier, that is, the part of the formula (to the right of the quantifier) to which the quantifier applies, may not be all that follows, in which case this is indicated by appropriate parentheses. For instance, in the universe of discourse $\mathbb{N}$, we may formalize the true statement that some integers are odd and that some are even by introducing predicates $O(x)$ for "x is odd" and $E(x)$ for "x is even" and write

$$(\exists x\, O(x)) \wedge (\exists x\, E(x)),$$

where the scope of each quantifier is indicated by parentheses, which allows for two independent uses of the variable x. This could be equivalently written

$$(\exists x\, O(x)) \wedge (\exists y\, E(y))$$

or, better yet,

$$\exists x \exists y\, O(x) \wedge E(y),$$

but **not** as

$$\exists x (O(x) \wedge E(x)),$$

which is a false statement, for no integer is both odd and even.

At any rate, quantifiers come before the predicate they quantify. For instance, one may be tempted to formalize a statement in plain English such as "every integer is less than some prime number," as follows (denoting by S the set of prime numbers[6]):

$$\forall n \in \mathbb{Z}\ n < p\ \exists p \in S.$$

That is NOT a statement with correct syntax. Here $\exists p \in S$ is a quantification that does not apply to any propositional function of the variable p. The correct way of reformulating this, first in English, then formally would be "for every integer, there is a prime number that is bigger":

---

[6]If we were to define this set formally, one may first say that given integers n and m, n *divides* m, in symbols $n|m$, if there is an integer k with $m = kn$:

$$n|m \iff \exists k \in \mathbb{Z}\, (m = kn)\,.$$

Then

$$S = \{p \in \mathbb{N} : (p > 1) \wedge \forall d \in \mathbb{N}\, (d|p \Longrightarrow (d = 1 \vee d = p))\}\,,$$

is the truth set (in the universe of discourse $\mathbb{Z}$) of

$$(p > 1) \wedge \forall d \in \mathbb{N}\, (d|p \Longrightarrow (d = 1 \vee d = p))\,.$$

$$\forall n \in \mathbb{Z}\ \exists p \in S\ (p > n), \tag{1.12}$$

which also turns out to be a true statement.

A **crucial point** is that a quantifier quantifies everything that follows so that the *order of quantification matters*. For instance, while (1.12) is a true statement, the statement

$$\exists p \in S\ \forall n \in \mathbb{Z}\ (p > n) \tag{1.13}$$

is false: there is no integer, in particular no prime number, that is greater than all integers. In (1.12), for each $n \in \mathbb{Z}$, there is a $p \in S$ *that depends on* n satisfying $p > n$. In contrast, the false statement (1.13) states the existence of a fixed $p \in S$ that, *independently of* $n \in \mathbb{Z}$, satisfies $p > n$.

1.3.4 Negating Quantified Statements

Most statements in Analysis involve multiple quantifiers. It is thus important to understand how to negate a quantified statement:

$$\neg\,(\forall x\, P(x)) \equiv \exists x\, \neg P(x) \tag{1.14}$$

for the statement "$P(x)$ is true for all x" is false if there is an x for which $P(x)$ is false, that is, if there is an x for which $\neg P(x)$ is true. Similarly

$$\neg\,(\exists x\, P(x)) \equiv \forall x\, \neg P(x), \tag{1.15}$$

for the statement "there is x for which $P(x)$ is true" is false if $P(x)$ is false for all x, that is, if $\neg P(x)$ is true for all x.

For instance, you may recall from Calculus that a function $f : \mathbb{R} \to \mathbb{R}$ is *continuous* at $x \in \mathbb{R}$ if

$$\forall \varepsilon > 0\, \exists \delta > 0\, \forall t \in \mathbb{R}\ (|x - t| < \delta \Longrightarrow |f(x) - f(t)| < \varepsilon)\,.$$

Thus a function is *discontinuous* at x if it is *not* continuous at x, that is, if

$$\begin{aligned}
&\neg\,(\forall \varepsilon > 0\, \exists \delta > 0\, \forall t \in \mathbb{R}\ (|x - t| < \delta \Longrightarrow |f(x) - f(t)| < \varepsilon)) \\
&\overset{(1.14)}{\equiv} \exists \varepsilon > 0\, \neg\,(\exists \delta > 0\, \forall t \in \mathbb{R}\ (|x - t| < \delta \Longrightarrow |f(x) - f(t)| < \varepsilon)) \\
&\overset{(1.15)}{\equiv} \exists \varepsilon > 0\, \forall \delta > 0\, \neg\,(\forall t \in \mathbb{R}\ (|x - t| < \delta \Longrightarrow |f(x) - f(t)| < \varepsilon)) \\
&\overset{(1.14)}{\equiv} \exists \varepsilon > 0\, \forall \delta > 0\, \exists t \in \mathbb{R}\, \neg\,(|x - t| < \delta \Longrightarrow |f(x) - f(t)| < \varepsilon) \\
&\overset{(1.5)}{\equiv} \exists \varepsilon > 0\, \forall \delta > 0\, \exists t \in \mathbb{R}\ (|x - t| < \delta \wedge |f(x) - f(t)| \geq \varepsilon)\,.
\end{aligned}$$

Note that in the last statement, ε is fixed, but *for each* $\delta > 0$, there is a t_δ that *depends on* δ that is within δ of x and whose image under f is at a distance greater or equal to ε from $f(x)$.

Exercise 1.56. If the universe of discourse is all countries,

1. introduce predicates to formalize the statement "some developed countries do not help underdeveloped countries"
2. spell out the form of its negation, and reformulate this negation in plain English.

Additional Exercises (Homework)

Exercise 1.57. Translate the following statements into quantified symbolic expressions. The universe of discourse is given in parentheses.

1. All precious stones are beautiful. (all stones)
2. Not all precious stones are beautiful. (all stones)
3. Some precious stones are not beautiful. (all stones)
4. Some stones are beautiful but not precious. (all stones)
5. Some isosceles triangles are equilateral. (all triangles)
6. Every right triangle is not isosceles. (all triangles)
7. Every triangle that is not isosceles is a right triangle. (all triangles)
8. No one knows everybody. (all people)
9. Everybody knows someone. (all people)

Exercise 1.58. If the universe of discourse is the set of all people, introduce appropriate predicates to give the logical form of the following statements:

1. If Batman is dangerous, then everybody is afraid.
2. Not everyone is both tired and bored.
3. Some criminals are neither armed nor violent.
4. Some soccer players dive or fake injuries.

Exercise 1.59. Which of the following (quantified) statements are true in the universe of all real numbers? Explain.

1. $\forall x\, \exists y\ (x+y=1)$
2. $\exists y\, \forall x\ (x+y=1)$
3. $\exists x\, \exists y\ \left(x^2+y^2=-4\right)$
4. $\forall x\, (x \neq 0 \Longrightarrow \exists y\, xy=1)$
5. $\forall x\, \exists y\, \forall z\ (xy=yz)$
6. $\exists x\, \forall y\ (x>y)$
7. $\forall y\, \exists x\ (x>y)$
8. $\exists! x\, \left(x^2=2\right)$
9. $\exists! x\, \left(x>0 \wedge x^2=2\right)$

Exercise 1.60. In the following, the actual meaning of "metric space," of "converge," and of "subsequence" is not important. Consider the following definition:

A metric space X is *compact* if every sequence on X has a subsequence that converges to some point of X.

1. Let X be a given metric space and $S(X)$ denote the set of sequences on X. Introduce appropriate predicates to formalize symbolically the statement that X is compact.
2. Write a useful formulation of "X is not compact."

Exercise 1.61. We revisit Exercise 1.43. If the universe of discourse is the set of players, introduce appropriate predicates and propositions to give the logical form of the statement "The players will go back to work if agreement is reached about their salaries, but this will be achieved, if at all, only if some of them take early retirement."

1.4 Some Basic Axioms of Set Theory

Now that we have developed the necessary basic logical language, we follow up on Section 1.1 and formulate some of the axioms of Set Theory. We do not develop axiomatic set theory here, and this section's purpose is simply to illustrate what sort of things are axiomatically required of a set, to complete the picture sketched in Section 1.1. ([7])

The usual axioms of Set Theory are those of Zermelo and Fraenkel (ZF), usually together with the Axiom of Choice (C), resulting in the ZFC system. As already pointed out, we will not study them systematically, but we may mention:

- that a set is determined by its elements via the *axiom of extensionality:*

$$X = Y \iff \forall z \, (z \in X \iff z \in Y).$$

- that the existence of an infinite set is axiomatic (*axiom of infinity*). This ensures in particular that at least one set exists.
- The *axiom of separation* states that for any predicate $\varphi(x)$ and every set X, the collection

$$\{x \in X : \varphi(x)\}$$

of elements of X satisfying $\varphi(x)$ is a set. Here the restriction to a set X is essential, and, in general, there is no set of all x satisfying $\varphi(x)$, unless x is restricted to a set to begin with:

[7] As mentioned in the introduction, the reader/instructor who prefers not to get into the axiomatic approach may choose to skip this section, and simply define equality of sets, then the empty set, and move on to the next section.

Example 1.62 (Russell's Paradox). Let $\varphi(x)$ denote the predicate $x \notin x$. Then

$$\{x : \varphi(x)\} = \{X : X \notin X\},$$

cannot be a set!

Assume to the contrary that $\{x : \varphi(x)\}$ defines a set Y. Then by definition,

$$Y \in Y \iff Y \notin Y.$$

As a consequence, there is no set of all sets, for otherwise, the collection above would be a set by the axiom of separation ([8]).

- that the existence of the *empty set* $\emptyset$, that is, a set with no element, follows from the axiom of infinity and that of separation. Indeed, if a set X exists (by the axiom of infinity), consider any predicate φ on X ([9]) and let

$$\emptyset = \{x \in X : \varphi(x) \wedge \neg\varphi(x)\}.$$

- the *axiom of pairing* ensures that given sets X and Y,

$$\{X, Y\}$$

is also a set. In particular, for $X = Y$, there is a set $\{X\}$ that has X as its only element. A set with exactly one element is called a *singleton*. This gives us a set $\emptyset$ with 0 element, and a set with 1 element: $\{\emptyset\}$. We can thus build a sequence

$$\emptyset, \{\emptyset\}, \{\emptyset, \{\emptyset\}\}, \{\emptyset, \{\emptyset\}, \{\emptyset, \{\emptyset\}\}\}, \ldots$$

of sets with $0, 1, 2, 3, \ldots$ elements, respectively ([10]). Induction (that we will study in detail in Section 2.6) produces sets of every finite cardinality ([11]).

- The axiom of *union* states that for every set X(of sets—for in set theory, everything is a set), there is a set

$$Y = \bigcup X = \bigcup_{A \in X} A$$

defined by

$$y \in \bigcup X \iff \exists A \in X \, (y \in A).$$

---

[8]In fact, another axiom called *axiom of regularity* ensures that no set is an element of itself, so that $\{X : X \notin X\}$ is the collection of all sets.

[9]Formally, as everything is a set, any x in X is also a set. Hence we may pick, for instance, $\varphi(x) =$"$x \in x$."

[10]To clarify the notation, note that, for instance, the set $\{\emptyset, \{\emptyset\}\}$ is the set whose two elements are the empty set and the set whose only element is the empty set.

[11]In fact, we may properly ground arithmetic in set theory by defining non-negative integers as particular sets:

$$0 = \emptyset;\ 1 = \{0\};\ 2 = \{0, 1\};\ 3 = \{0, 1, 2\}; \ldots \tag{1.16}$$

Even though we do not think of $0, 1, 2, \ldots$ as sets, we can define them as such to make arithmetic part of set theory. It is in this sense that set theory provides a foundation for all mathematics.

- The *Axiom of Choice* (AC) is not as widely accepted. It is in particular contested by constructivist mathematicians, who take issue with existence results that are not based on an explicit construction. (AC) states that for any set X of non-empty sets, there exists a *choice function* $f : X \to \bigcup X$, that is, a function $f : X \to \bigcup X$ satisfying

$$\forall A \in X \; (f(A) \in A).$$

In other words, a choice function picks an element out of each set $A \in X$. It is a benign operation for a finite collection of sets. That we can make a simultaneous choice of an element out of each set of an *infinite* collection, without having a rule to make this choice ([12]), is what requires an axiom, and is more controversial because of some of its consequences. However, we will take (AC) for granted, as most mathematicians do.

1.5 Subsets

Definition 1.63. A set A is a *subset* of a set X, in symbols

$$A \subset X,$$

if every element of A is an element of X, that is, if

$$x \in A \Longrightarrow x \in X.$$

In other words,

$$A \subset X \iff \forall x (x \in A \Longrightarrow x \in X). \tag{1.17}$$

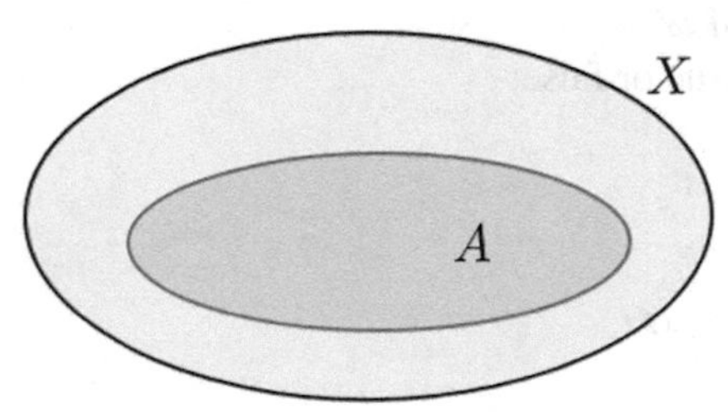

[12] If, for instance, each set $A \in X$ had a smallest element, that would provide a rule to make a choice of an element of each A, and (AC) would not be needed to obtain a choice function.

Bertrand Russell illustrated this point with an example involving shoes and socks. Imagine a drawer that contains infinitely many pairs of shoes. How do you pick a shoe from each pair? What if the drawer contains infinitely many pairs of socks? In the first case, you can always pick the right shoe, and (AC) is not required to make this choice, because a definite way to pick is available. In contrast, socks of the same pair are indistinguishable, and therefore, no rule to pick one rather than the other can be applied systematically. Hence (AC) is here needed to form a set by picking one sock out of each pair.

Definition 1.64. The collection of all subsets of a given set X is a set (this is also an axiom!), called the *powerset of* X and denoted $\mathbb{P}X$ or sometimes 2^X. Hence

$$Y \in \mathbb{P}X \iff Y \subset X.$$

One should pay particular attention to distinguish the relations $\subset$ and $\in$! For instance, while $\emptyset$ does not need to be an element of a given set, it is a subset of every set:

Proposition 1.65. *If X is a set, then $\emptyset \subset X$ and $X \subset X$.*

Proof. Recall that a conditional statement $p \Longrightarrow q$ is true whenever the premise p is false (see Section 1.2). In view of (1.17), $x \in \emptyset \Longrightarrow x \in X$ is true, because $x \in \emptyset$ is false by definition of $\emptyset$. Similarly, $x \in X \Longrightarrow x \in X$ is always true. □

A *non-empty* subset of X that is not equal to X is called a *proper subset of* X.

Remark 1.66. Some authors use $\subseteq$ for the inclusion relation and $\subset$ or $\subsetneq$ to emphasize that the subset is not the whole set. We do not. We will not use the symbol $\subseteq$ at all, and if you encounter it somewhere else, remember that it is exactly our symbol $\subset$.

Exercise 1.67. List all subsets (proper or not) of $\{1\}$, of $\{\alpha, \beta\}$, of $\{1,2,3\}$ and of $\{a,b,c,d\}$. How many are there?

We will see in Section 1.7 a general result (Corollary 1.116) on the number of subsets of a finite set.

To clarify further the distinction between membership and inclusion, consider the following:

Exercise 1.68. Let $\mathscr{A}$ be the set of all subsets of $X = \{1,2,3,4\}$ with 2 elements.

1. List the elements of $\mathscr{A}$;
2. Are the following true or false:

   a. $1 \in \mathscr{A}$;
   b. $\{1,2\} \in \mathscr{A}$;
   c. $\{1,2\} \subset \mathscr{A}$;
   d. $\{\{1,2\},\{2,3\}\} \subset \mathscr{A}$;
   e. $\{\{3,4\}\} \subset \mathscr{A}$;
   f. $\emptyset \in \mathscr{A}$;
   g. $\emptyset \subset \mathscr{A}$.

Note that $A \subset B$ and $B \subset A$ means $(x \in A \Longrightarrow x \in B)$ and $(x \in B \Longrightarrow x \in A)$, that is,

$$x \in A \iff x \in B,$$

so that

$$A = B \iff (A \subset B) \wedge (B \subset A). \tag{1.18}$$

This is very often what we use to prove the equality between two sets: by showing two inclusions.

Exercise 1.69. Show that if $A \subset B$ and $B \subset C$ then $A \subset C$, that is, $\subset$ is a transitive relation.

In many cases, all sets considered are subsets of a fixed set, often called *ambient set* or *universal set*. The ambient set depends of course on the context.

Note that (1.18) is reminiscent of (1.7) and Exercise 1.69 is reminiscent of Exercise 1.29. In fact, if X is the universe of discourse for your predicates, and we take X as the ambient set, then *inclusion models implication* ([13]). Namely, if $p(x)$ is a predicate, then its truth set is the set

$$P = \{x \in X : p(x)\}$$

of values of $x \in X$ for which $p(x)$ is true. If Q is the truth set of $q(x)$, then the implication

$$\forall x\,(p(x) \Longrightarrow q(x))$$

is equivalent to the inclusion

$$P \subset Q,$$

so that the set-theoretic form $P \subset Q$ models the logical form $p \Longrightarrow q$.

More generally logical connectives find set-theoretic interpretations in terms of operations on sets, as we will see in the next section.

Exercise 1.70. Show that

$$A \subset B \Longleftrightarrow \mathbb{P}(A) \subset \mathbb{P}(B).$$

Additional Exercises (Homework)

Exercise 1.71. What is the truth value of each of the following statements:

1. $\{2,3,7\} \subset \{2,4,5,6,7\}$;
2. $\{2n : n \in \mathbb{N}\} \subset \mathbb{N}$;
3. $\emptyset \notin \mathbb{N}$;
4. $\emptyset \subset \{1,2\}$;
5. $\emptyset \in \{1,2\}$;
6. $\emptyset \in \mathbb{P}(\{1,2\})$;
7. $\emptyset \subset \mathbb{P}(\{1,2\})$;
8. $1 \in \mathbb{P}(\{1,2\})$;
9. $\{2\} \subset \mathbb{P}(\{1,2\})$;
10. $\{2\} \in \mathbb{P}(\{1,2\})$;
11. $\emptyset \in \{\{\emptyset\}\}$;
12. $\{2\} \subset \{1,\{2\}\}$;

[13] Here "models" is not meant to have a technical meaning, but rather to underline the tight connection between the two notions. One may alternatively say that inclusion is the set-theoretic formulation of implication.

13. $\{1,2\} \in \{1,2\}$;
14. $\{1,2\} \subset \{1,2\}$;
15. $7 \in \mathbb{N}$;
16. $7 \subset \mathbb{N}$;
17. $\{7\} \subset \mathbb{N}$;
18. $\{7\} \in \mathbb{P}(\mathbb{N})$;
19. $\{7\} \subset \mathbb{P}(\mathbb{N})$;
20. $\{\{7\}\} \subset \mathbb{P}(\mathbb{N})$;
21. $\{\{1\}\} \in \mathbb{P}(\{1,\{1\}\})$;
22. $\{\{1\},\{\{1\}\}\} \subset \mathbb{P}(\{1,\{1\}\})$.

Exercise 1.72. Show that if $A \subset B$, $B \subset C$, and $C \subset A$, then $A = B = C$.

Exercise 1.73. Is there only one empty set? Justify your answer.

Exercise 1.74. Given $a \in \mathbb{N}$, let $a\mathbb{N} = \{n \in \mathbb{N} : \exists k \in \mathbb{N}\, n = ka\}$. Show that $4\mathbb{N} \subset 2\mathbb{N}$ and that $2\mathbb{N} \not\subset 4\mathbb{N}$.

Exercise 1.75. Write out explicitly these sets (giving their list of elements):

1. $\mathbb{P}(\{\{1,2\},\{3\}\})$;
2. $\mathbb{P}(\mathbb{P}(\{\emptyset\}))$.

1.6 Operations on Sets and Logical Connectives

Definition 1.76. The *set-theoretic difference* is defined as:

$$A \setminus B = \{x \in A : x \notin B\}.$$

When there is a specific ambient set X, the difference $X \setminus A$ is called *complement of* A (in X):

$$A^c = X \setminus A = \{x \in X : x \notin A\}.$$

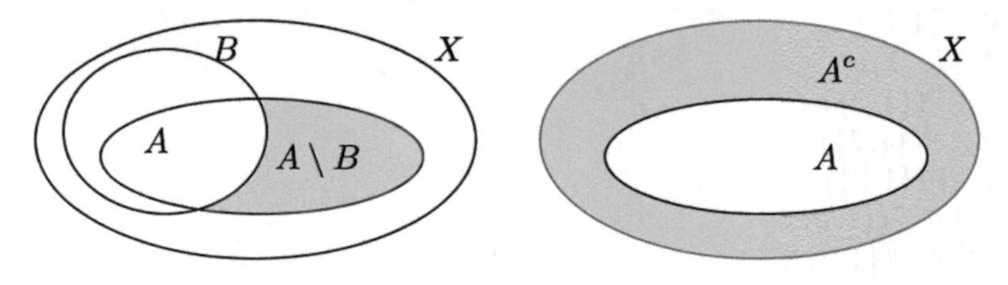

Complementation models negation: if P is the truth set of $p(x)$, then P^c is the truth set of $\neg p(x)$.

Note that $\neg$T$\equiv$ F is modeled by

$$X^c = \emptyset,$$

and ¬F≡ T by

$$\emptyset^c = X,$$

for X is the truth set for a propositional form that is always true while the empty set is the truth set for one that is always false. Double negation is identity, that is, for any proposition p,

$$\neg(\neg p) \iff p,$$

which in set-theoretic terms means that for any subset A of X,

$$(A^c)^c = A.$$

Since inclusion models implication and complementation models negation, the observation (1.6) that a conditional statement is equivalent to its contrapositive rephrases in set-theoretic terms as, for every subsets A, B of X:

$$A \subset B \iff B^c \subset A^c.$$

In view of (1.18),

$$A = B \iff A^c = B^c.$$

Definition 1.77. The *intersection* $A \cap B$ of sets A and B is the set of elements that are in both A and B:

$$A \cap B = \{x : x \in A \wedge x \in B\};$$

The *union* $A \cup B$ of two sets A and B is the set of elements that belong to at least one of the sets:

$$A \cup B = \{x : x \in A \vee x \in B\}.$$

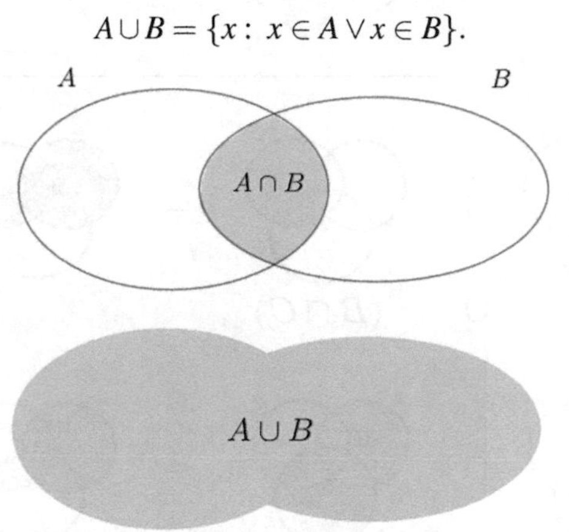

Intersection models conjunction and union models disjunction: If P and Q are the truth sets of $p(x)$ and $q(x)$, then $P \cap Q$ is the truth set of $p(x) \wedge q(x)$ and $P \cup Q$ is the truth set of $p(x) \vee q(x)$.

As a consequence, the logical associative laws on page 7 rephrase in set-theoretic terms as

$$A \cap (B \cap C) = (A \cap B) \cap C \text{ and } A \cup (B \cup C) = (A \cup B) \cup C. \tag{1.19}$$

Distributive laws for logical connectives (Proposition 1.19) rephrases in the set-theoretic context as:

Proposition 1.78 (Distributive Laws). *Let A, B, and C be sets. Then*

$$A\cup(B\cap C) = (A\cup B)\cap(A\cup C) \tag{1.20}$$

$$A\cap(B\cup C) = (A\cap B)\cup(A\cap C). \tag{1.21}$$

The Venn diagrams below illustrate the distributive laws. The first two rows illustrate the left- and right-hand sides, respectively of (1.20), and the next two illustrate (1.21) in a similar fashion.

$A \cap (B\cup C)$

$(A\cap B) \cup (A\cap C)$

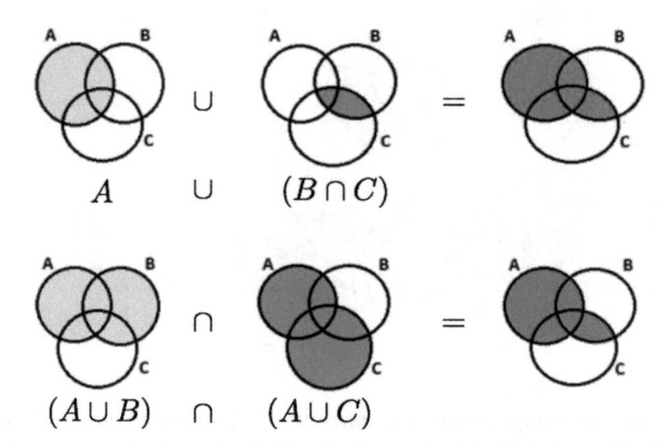

Similarly, since $\cup$ models $\vee$, $\cap$ models $\wedge$ and complement models negation, the DeMorgan laws of logic (Proposition 1.15) find their set-theoretic counterpart:

Proposition 1.79 (De Morgan's Laws (Set-Theoretic Form)). *If A and B are subsets of a universal set X then*

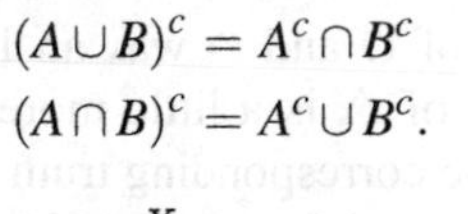

$$(A \cup B)^c = A^c \cap B^c$$
$$(A \cap B)^c = A^c \cup B^c.$$

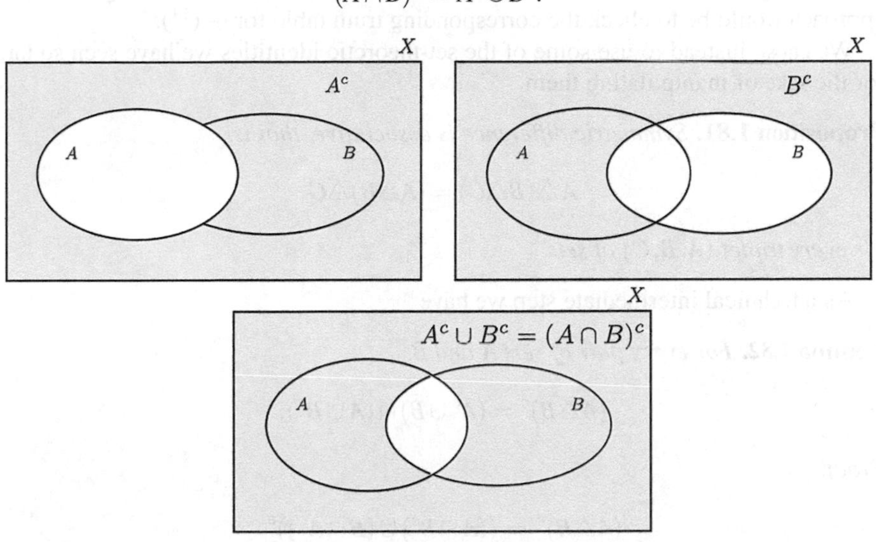

Definition 1.80. The *symmetric difference* $A \triangle B$ of subsets A and B of X is the set of points that are in one set and not the other. In other words,

$$\begin{aligned} A \triangle B &= (A \setminus B) \cup (B \setminus A) \\ &= (A \cap B^c) \cup (B \cap A^c) \\ &= (A \cup B) \setminus (A \cap B). \end{aligned}$$

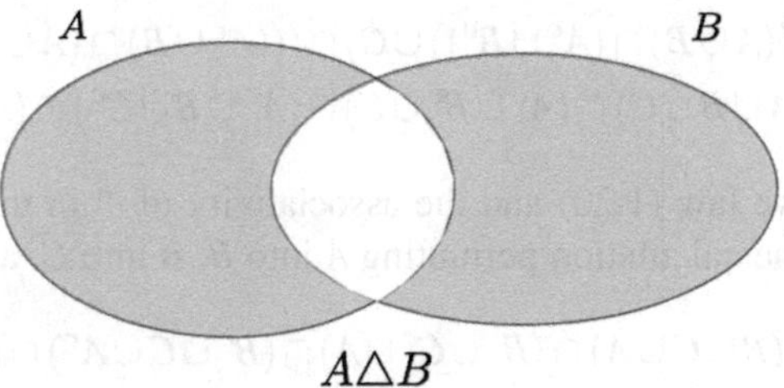

Symmetric difference models "exclusive or" (sometimes denoted XOR or $\oplus$), that is, the truth set of $p(x) \oplus q(x)$ (where $p \oplus q$ is true when one or the other, but not both, of p and q are true) is $P \triangle Q$.

While associativity (1.19) of $\cup$ and $\cap$ was an immediate consequence of the associativity of $\vee$ and $\wedge$, that of $\triangle$ is a little more delicate. One straightforward approach would be to check the corresponding truth table for $\oplus$ ([14]).

We chose instead to use some of the set-theoretic identities we have seen so far, for the sake of manipulating them.

Proposition 1.81. *Symmetric difference is associative, that is,*

$$A\triangle(B\triangle C) = (A\triangle B)\triangle C$$

for every triplet (A, B, C) of sets.

As a technical intermediate step we have

Lemma 1.82. *For every pair of sets A and B,*

$$(A\triangle B)^c = (A^c \cup B) \cap (A \cup B^c).$$

Proof.

$$\begin{aligned}(A\triangle B)^c &= ((A\cap B^c)\cup(B\cap A^c))^c \\ &= (A\cap B^c)^c \cap (B\cap A^c)^c \\ &= (A^c\cup B)\cap(B^c\cup A),\end{aligned}$$

using De Morgan's laws twice. □

Proof (Proof of Proposition 1.81). Now

$$\begin{aligned}(A\triangle B)\triangle C &= ((A\triangle B)\cup C)\cap((A\triangle B)^c\cup C^c) \\ &= (((A\cup B)\cap(A^c\cup B^c))\cup C)\cap(((A^c\cup B)\cap(A\cup B^c))\cup C^c) \\ &= (A\cup B\cup C)\cap(A^c\cup B^c\cup C)\cap(A^c\cup B\cup C^c)\cap(A\cup B^c\cup C^c), \quad (*)\end{aligned}$$

using the distributive law (1.20) and the associativity of $\cap$ in the last identity. Observing that the same calculation permuting A into B, B into C, and C into A yields

$$(B\triangle C)\triangle A = (B\cup C\cup A)\cap(B^c\cup C^c\cup A)\cap(B^c\cup C\cup A^c)\cap(B\cup C^c\cup A^c),$$

[14]namely,

p	q	r	$p\oplus q$	$(p\oplus q)\oplus r$	$(q\oplus r)$	$p\oplus(q\oplus r)$
T	T	T	F	T	F	T
T	T	F	F	F	T	F
T	F	T	T	F	T	F
T	F	F	T	T	F	T
F	T	T	T	F	F	F
F	T	F	T	T	T	T
F	F	T	F	T	T	T
F	F	F	F	F	F	F

where the right-hand side is equal to that of (*), we obtain

$$(A \triangle B) \triangle C = (B \triangle C) \triangle A = A \triangle (B \triangle C).$$

□

Exercise 1.83. Show that the powerset $\mathbb{P}X$ of a set X, equipped with the symmetric difference $\triangle$ satisfies the following:

1.
$$\exists E \in \mathbb{P}X, \forall A \in \mathbb{P}X,\ A \triangle E = E \triangle A = A;$$

2. For the set E of the previous question
$$\forall A \in \mathbb{P}X, \exists B \in \mathbb{P}X : A \triangle B = E.$$

Solution. Let $E = \emptyset$. It is easily verified that $A \triangle \emptyset = A$ for every $A \in \mathbb{P}X$. On the other hand, $A \triangle A = \emptyset$ for every $A \in \mathbb{P}X$.

Remark 1.84. Note that $\triangle$ is a commutative binary operation on $\mathbb{P}X$ (that is, it takes two elements of $\mathbb{P}X$ and returns an element of $\mathbb{P}X$). Proposition 1.81 shows that this operation is also associative. The first question of Exercise 1.83 shows that $\emptyset$ is a *neutral element*: it does not change other sets under this operation. The second question shows that every $A \in \mathbb{P}X$ has *an inverse* (itself!) for the operation $\triangle$: a set that when "triangled" by A returns the neutral element. This is very much the same structure that you observe in say $(\mathbb{Z}, +)$: $+$ is a commutative and associative binary operation on $\mathbb{Z}$ with a neutral element 0 (for all $n \in \mathbb{Z}$, $n+0 = 0+n = n$), and every element has an inverse (for all $n \in \mathbb{Z}$, the inverse $-n$ satisfies $n + (-n) = 0$). These are two examples of a (commutative) *group*.

Mathematicians like to identify ubiquitous patterns, or *structures*, and introduce the corresponding concept (like that of group) abstractly to develop a theory that applies to large classes of objects. To a large extent branches of mathematics correspond to the study of various kinds of such structures.

Of course, given a set $\mathscr{A}$ of sets, we can define its union as in the axiom of union

$$\bigcup \mathscr{A} = \bigcup_{A \in \mathscr{A}} A = \{x : \exists A \in \mathscr{A}\ x \in A\},$$

and its *intersection* similarly:

$$\bigcap \mathscr{A} = \bigcap_{A \in \mathscr{A}} A = \{x : \forall A \in \mathscr{A}\ x \in A\}.$$

Exercise 1.85. Given an example of a sequence $\{A_n\}_{n=1}^{\infty}$ of decreasing subsets of $\mathbb{N}$ (that is, $A_{n+1} \subset A_n \subset \mathbb{N}$ for every $n \in \mathbb{N}$) such that

$$\bigcap_{n=1}^{\infty} A_n = \emptyset \text{ and } \bigcup_{n=1}^{\infty} A_n = \mathbb{N}.$$

Definition 1.86. A set $\mathscr{A} \subset \mathbb{P}X$ of subsets of X is *pairwise disjoint* if $A_1 \cap A_2 = \emptyset$ for every $A_1 \in \mathscr{A}$ and $A_2 \in \mathscr{A}$.

Example 1.87. For instance,

$$\mathscr{A} = \{\{1,3,5\},\{2,4,6\},\{\square,\triangle\}\}$$

is pairwise disjoint because $\{1,3,5\} \cap \{2,4,6\} = \emptyset$, $\{1,3,5\} \cap \{\square,\triangle\} = \emptyset$ and $\{2,4,6\} \cap \{\square,\triangle\} = \emptyset$, while

$$\mathscr{B} = \{\{1,3,5\},\{2,4,6\},\{1,7,8\}\}$$

is not pairwise disjoint because $\{1,3,5\} \cap \{1,7,8\} = \{1\} \neq \emptyset$. Note however that $\bigcap \mathscr{B} = \emptyset$, for there is no element that is common to all three sets in $\mathscr{B}$.

De Morgan's laws are actually valid also in the infinite case:

Exercise 1.88. Let $\mathscr{A} \subset \mathbb{P}X$, that is, elements of $\mathscr{A}$ are subsets of X. Show that

$$\left(\bigcap_{A \in \mathscr{A}} A\right)^c = \bigcup_{A \in \mathscr{A}} A^c$$
$$\left(\bigcup_{A \in \mathscr{A}} A\right)^c = \bigcap_{A \in \mathscr{A}} A^c.$$

On the other hand, there are infinite distributivity laws, called *frame law* and *coframe law*, respectively, that generalizes (1.21) and (1.20):

Exercise 1.89. Show that if $\mathscr{B} \subset \mathbb{P}X$ and $A \in \mathbb{P}X$ then

$$A \cap \bigcup_{B \in \mathscr{B}} B = \bigcup_{B \in \mathscr{B}} (A \cap B)$$

and

$$A \cup \bigcap_{B \in \mathscr{B}} B = \bigcap_{B \in \mathscr{B}} (A \cup B).$$

Additional Exercises (Homework)

Exercise 1.90. Let $A = \{1,2,3,5,8,9\}$, $B = \{0,2,4,6,8\}$, $C = \{1,2,3,4\}$, and $D = \{1,2,3,7,9,10\}$. Find:

1. $A \cup B$;
2. $A \cap B$ and $A \setminus B$;
3. $C \setminus B$ and $C \setminus A$;
4. $A \triangle C$ and $A \triangle D$;
5. $(A \cup C) \cap (B \setminus D)$;
6. $D \cap (B \cup C)$.

Exercise 1.91. Let $X = (-\infty, 3] \cup [7, 10)$ and let $A = [-10, 9] \cap X$. What is the complement of A in X?

Exercise 1.92. Give an example of distinct sets A, B, and C such that

1. $C \subset A \cup B$ but $A \cap B \not\subset C$;
2. $A \cup B \subset C$ and $C \not\subset B$ and $C \not\subset A$;
3. $A \subset B \cup C, B \subset A \cup C, C \subset A \cup B$;
4. $A \cap B \subset C, A \cap C \subset B, B \cap C \subset A$ and $A = B \cup C$.

Exercise 1.93. Consider the following intervals of $\mathbb{R}$: $A = [2,5]$, $B = [3,7]$, $C = [2,3]$, $D = (3, \infty)$. Describe

1. $A \cup B$, $A \cup C$ and $A \cup D$;
2. $A \cap B$, $A \cap C$, $A \cap D$, $B \cap D$ and $C \cap D$;
3. $\mathbb{R} \setminus A$ and $\mathbb{R} \setminus D$;
4. $A \setminus B$, $B \setminus A$ and $A \triangle B$;
5. $B \triangle D$.

Exercise 1.94. Let $A = \{1,2,3,4\}$, $B = \{2,5,7\}$, $C = \{a,b,c,d,e\}$ and $D = \{b,c,f,g,h\}$. Describe explicitly the following sets:

1. $A \cap B$, $A \cup B$ and $A \triangle B$;
2. $C \cap D$, $C \setminus D$, $C \cup D$ and $C \triangle D$;
3. $\mathbb{P}(C \cap D)$.

Exercise 1.95. Provide counterexamples ([15]) to each of the following:

1.
$$X \cup Z \subset Y \cup Z \Longrightarrow X \subset Y.$$

2.
$$(X \setminus Y) \cap (X \setminus Z) = \emptyset \Longrightarrow Y \cap Z = \emptyset.$$

3.
$$\mathbb{P}(X) \setminus \mathbb{P}(Y) \subset \mathbb{P}(X \setminus Y).$$

Exercise 1.96. Find the union and intersection of the following families of sets:

1. $\mathscr{A} = \{\{1,2,3\}, \{3,7,10\}, \{2,3,5,7,9,10\}, \{3,10,15\}\}$.
2.
$$\mathscr{A} = \left\{ \left(0, \frac{3}{n}\right) : n \in \mathbb{N} \right\}.$$

3.
$$\mathscr{A} = \{[1,r) : r \in (1,5)\}.$$

---

[15] That is, examples of sets X, Y, and Z showing that the statement may be false. See Section 2.5 for more details on counterexamples.

4. For each $p \in \mathbb{N}$, let $p\mathbb{N} = \{pn : n \in \mathbb{N}\}$ denote the set of multiples of p. Let

$$\mathscr{A} = \{p\mathbb{N} : p \text{ prime}\}.$$

Exercise 1.97. Find a family $\mathscr{A}$ of four subsets of $X = \{a,b,c,d,e,f,g,h\}$ with $\bigcap \mathscr{A} = \{a,c\}$ and $\bigcup \mathscr{A} = X$. Find another family $\mathscr{B}$ of four subsets of X that is pairwise disjoint and satisfies $\bigcup \mathscr{B} = X$.

1.7 Functions and Counting: First Look

We make a detour through the notion of *function*, in part to be able to count subsets of a finite set. We will revisit the notion of function in more detail in Chapter 3. Recall that

Definition 1.98. If X and Y are sets, a *function* or *map* $f : X \to Y$ associates with each $x \in X$ exactly one element $f(x)$ of Y. The set X is the *domain* of f, while Y is its *codomain.* A function is *one-to-one* (or *injective*) if it doesn't take the same value twice, that is, if

$$\forall x,t \in X\,(f(x) = f(t) \Longrightarrow x = t),$$

and *onto* (or *surjective*) if

$$\forall y \in Y\, \exists x \in X\,(f(x) = y).$$

A *one-to-one correspondence* or *bijection* is a map that is both one-to-one and onto. We may also say that a map is *bijective* to say that it is a bijection.

Remark 1.99. Saying that a function *associates* with each element of the domain exactly one element of the codomain is not quite as precise as one would wish, for the term "associates" is rather vague. The exact nature of the association will be clarified in Remark 1.135, when we are in a position to introduce the graph of a function.

Definition 1.100. Given a set X, we denote by $i_X : X \to X$ the *identity function* defined by $i_X(x) = x$ for all $x \in X$.

Of course, identity functions are bijections.

Example 1.101. The map $f : \mathbb{R} \to \mathbb{R}$ defined by $f(x) = x^2$ is not one-to-one because $f(-1) = f(1)$. It is not onto because f takes only non-negative values. Hence, -1 is not the image of any real number under f. On the other hand, the map $g : [0,\infty) \to \mathbb{R}$ defined by $g(x) = x^2$ is one-to-one but not onto. Indeed, if $g(x) = g(t)$, that is $x^2 = t^2$ with $x,t \in [0,\infty)$, then $x = t$. That g is not onto is for the same reason that f is not onto. Now $h : \mathbb{R} \to [0,\infty)$ defined by $h(x) = x^2$ is not one-to-one (for the same reason that f is not) but it is onto, because every $y \geq 0$ is the square of a real number, for instance $\sqrt{y}$. Finally, $\ell : [0,\infty) \to [0,\infty)$ defined by $\ell(x) = x^2$ is a bijection.

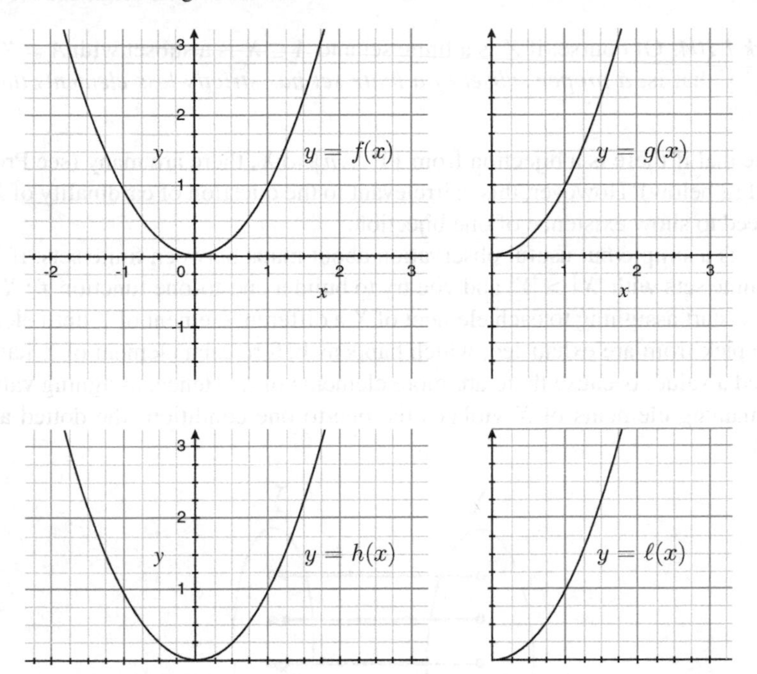

Recall as well that if $f: X \to Y$ and $g: Y \to Z$, then the *composite function* $g \circ f: X \to Z$ is defined by $g \circ f(x) = g(f(x))$.

It is not hard to prove (see Exercise 2.13, which you may attempt now, or after we discuss proof structure further) that:

Proposition 1.102. *Let $f: X \to Y$ and $g: Y \to Z$.*

1. *If f and g are one-to-one, so is $g \circ f$.*
2. *If f and g are onto, so is $g \circ f$.*
3. *If f and g are bijections, so is $g \circ f$.*
4. *If $g \circ f$ is onto, so is g.*
5. *If $g \circ f$ is one-to-one, so is f.*

Remark 1.103. When a child learns to count on her fingers, say till 5, she "counts" members of a collection of up to 5 objects by associating each object to one of her fingers. She counts *each* object, but doesn't count *any twice*. In other words, she builds an *onto* (every object is counted) and *one-to-one* (objects are counted only once) function, hence a *bijection*, from a subset of her fingers to the set of counted objects. The number of objects is then the number of fingers, which is implicitly in bijection with $\{1,2,\ldots,n\}$ where here $n \leq 5$.

In other words, a finite set X has *cardinality* n if there is a bijection

$$f: \{1,2,\ldots n\} \to X.$$

We then write $|X| = n$, that is, $|X|$ stands for the cardinality of X.

Remark 1.104. Of course, if X is a finite set and $A \subset X$ is a subset with $A \neq X$, then $|A| < |X|$, that is, *a proper subset of a finite set has strictly less elements than the whole set.*

Note that if there is a bijection from $\{1, \ldots n\}$ to X, there are many (see Proposition 1.112 below). However, this is irrelevant to the question of cardinality of X. We only need to show existence of one bijection.

Here is a simple but useful observation about maps between finite sets: if X and Y are finite sets with $|X| > |Y|$ and you try to build a one-to-one function $f : X \to Y$, then you start assigning to each element of X a different element of Y until elements of Y to pick from are exhausted, which happens before each element of X has been assigned a value, because there are more elements of X. Hence, assigning values to the remaining elements of X violates the one-to-one condition (the dotted arrows below):

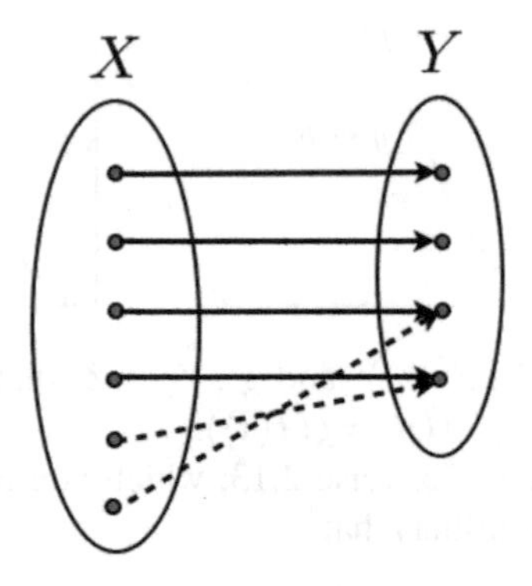

Another way to think of this situation is that if X is a set of pigeons and Y a set of pigeonholes, then if there are more pigeons than pigeonholes and all pigeons enter a pigeonhole, then at least one hole will contain more than one pigeon. Hence this simple observation is often called the *Pigeonhole Principle.*

Similarly, if $|Y| > |X|$ and you try to build an onto function from X to Y, you also assign to each element of X a different element of Y to maximize the number of elements of Y that you obtain as an image of an element of X. But obviously, you can only get $|X|$ of them, that is, the map cannot be onto.

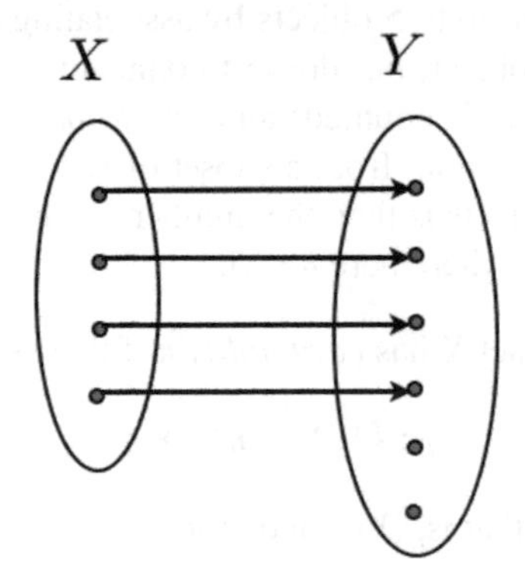

In other words:

Proposition 1.105 (Pigeonhole Principle). *Let $f : X \to Y$ where X and Y are finite sets*

1. *If $|X| > |Y|$, then f is not one-to-one;*
2. *If $|Y| > |X|$, then f is not onto.*

Rephrasing in terms of the contrapositives, we obtain:

Corollary 1.106. *Let $f : X \to Y$ where X and Y are finite sets*
1. *If f is one-to-one, then $|X| \leq |Y|$;*
2. *If f is onto, then $|Y| \leq |X|$.*

While the Pigeonhole Principle is a very simple observation, its consequences are not always intuitive:

Example 1.107. In a large lecture hall of 400 students, at least two students share the same first name initial and the same month of birth, and at least two students share the same birthday. Indeed, there are 26 possible initials and 12 possible months of birth, hence there are $26 \times 12 = 312$ possible pairs. Hence the map associating to each student the pair (I, M) where I is the initial of her first name and M her birth month is not one-to-one, because the domain has more elements than the co-domain. In other words, at least two students have the same image. Similarly, there are only 366 possible birthdays, and 400 students, so at least 2 students have the same birthday.

Remark 1.108. In fact, we can say a little more: if $f : X \to Y$ and $y \in Y$, let

$$f^{-1}[y] = \{x \in X : f(x) = y\}, \tag{1.22}$$

denote the *fiber of y under f*. If $|X| > |Y|$, then there is $y \in Y$ with

$$|f^{-1}[y]| \geq \left\lceil \frac{|X|}{|Y|} \right\rceil, \tag{1.23}$$

where $\lceil r \rceil$ denotes the least integer greater or equal to r.

Otherwise, there would be $|Y|$ fibers, each with less than $\frac{|X|}{|Y|}$ elements (because each fiber has a cardinality that is an integer), and thus X would have less than $|X|$ elements, which is not possible. Hence, for example, to guarantee that a group of n individuals contains 3 people with the same birthday, n should be such that

$$\left\lceil \frac{n}{366} \right\rceil \geq 3.$$

The smallest such integer is

$$n = 366 \times 2 + 1 = 733.$$

Note that some of the notions of Definition 1.98 can be easily rephrased in terms of fibers, as defined by (1.22). Namely:

Proposition 1.109. *A function $f : X \to Y$ is*

1. *onto if and only if $f^{-1}[y] \neq \emptyset$ for every $y \in Y$;*
2. *one-to-one if and only if $|f^{-1}[y]| \leq 1$ for every $y \in Y$;*
3. *a bijection if and only if $|f^{-1}[y]| = 1$ for every $y \in Y$.*

In particular, if one seeks a function $g : Y \to X$ that "undoes what f does," that is, satisfying $g(f(x)) = x$ for all $x \in X$, then we need to have a unique x to pick in the fiber $f^{-1}[f(x)]$ as the value $g(y)$ where $y = f(x)$. Hence, we need f to be one-to-one. If, moreover, we want $f(g(y)) = y$ for all $y \in Y$, then we need of course f to be onto. More formally:

Definition 1.110. If $f : X \to Y$ is a function, an *inverse function* for f is a function $g : Y \to X$ with

$$g \circ f = i_X \text{ and } f \circ g = i_Y.$$

Proposition 1.111. *A function f admits an inverse function if and only if f is a bijection. In this case, the inverse function is unique and a bijection.*

Proof. First note that if $f : X \to Y$ has an inverse function g, then f is one-to-one. Indeed, if $x_1, x_2 \in X$ satisfy $f(x_1) = f(x_2)$, then

$$x_1 = g(f(x_1)) = g(f(x_2)) = x_2,$$

and thus f is one-to-one. Moreover, for each $y \in Y$, we have $y = f(g(y))$ so that f is onto.

If $f : X \to Y$ had two inverse maps $g_1, g_2 : Y \to X$, then for every $y \in Y$,

$$f(g_1(y)) = y = f(g_2(y))$$

and f is one-to-one, so that $g_1(y) = g_2(y)$. Hence $g_1 = g_2$.

If f is a bijection, for every $y \in Y$, there is a unique x_y in the fiber $f^{-1}[y]$ and we can thus define a function $g : Y \to X$ by $g(y) = x_y$. By definition, $f \circ g = i_Y$ and $g \circ f = i_X$. □

We denote by

$$f^{-1} : Y \to X$$

the (unique) inverse function of f. With these notations, a bijection $f : X \to Y$ has an inverse function $f^{-1} : Y \to X$ and $f^{-1}[y] = \{f^{-1}(y)\}$. As $f \circ f^{-1} = i_Y$ and $f^{-1} \circ f = i_X$, the function f^{-1} admits f as an inverse and is thus a bijection, and $(f^{-1})^{-1} = f$.

As stated before, when there is a bijection between two finite sets, there are many. Namely,

Proposition 1.112. *If*

1. *X is a non-empty set with n elements, there are*

$$n! = 1 \times 2 \times 3 \times \ldots \times n$$

bijections $f : \{1, \ldots n\} \to X$. In particular, there are $n!$ permutations of objects of X.

2. *$1 \le k \le n$, $|X| = k$ and $|Y| = n$, then there are*

$$n(n-1)\ldots(n-k+1) = \frac{n!}{(n-k)!}$$

one-to-one maps $f : X \to Y$.

Proof. (1) There are n choices to pick $f(1)$. To pick $f(2)$, any element of X can be chosen, except $f(1)$, to keep the map one-to-one. Hence there are $n-1$ choices for $f(2)$ and thus $n(n-1)$ choices for the pair $\{f(1), f(2)\}$. Similarly, there are $n-2$ choices for $f(3)$ and thus $n(n-1)(n-2)$ choices for the triplet $\{f(1), f(2), f(3)\}$. Proceeding this way, we have

$$n(n-1)(n-2)\ldots 2$$

ways to assign different values of X to 1 through $n-1$, and only one choice left for $f(n)$. Hence, there are $n!$ ways to assign the values of the elements of $\{1 \ldots n\}$ under f for a one-to-one map, which is then also onto by construction. In other words, there are $n!$ bijections.

(2) This is the same argument, except that we assign the last value $f(x_k)$ potentially before the n possible values in Y are exhausted. □

Remark 1.113. To give explicitly the number of onto maps between two finite sets is more complicated, and the argument would not be very palatable at this point, before you develop some experience with proofs. We include this in the complements in Appendix A (Corollary A.6).

We will now show that subsets of X are in one-to-one correspondence with functions from X to $\{0, 1\}$ (Proposition 1.114 below). Indeed, with a subset A of X, we can associate its *indicator function* $\chi_A : X \to \{0, 1\}$ defined by

$$\chi_A(x) = \begin{cases} 1 & \text{if } x \in A \\ 0 & \text{if } x \notin A \end{cases}.$$

Denoting by $\{0, 1\}^X$ the set of all functions from X to $\{0, 1\}$, we have:

Proposition 1.114. *The map $F : \mathbb{P}X \to \{0, 1\}^X$ defined by $F(A) = \chi_A$ is a bijection.*

Proof. This function is one-to-one for if $F(A) = F(B)$, that is, $\chi_A = \chi_B$, then $\{x \in X : \chi_A(x) = 1\} = A$ and $\{x \in X : \chi_B(x) = 1\} = B$ coincide. F is also onto, for if $f : X \to \{0, 1\}$ then $f = \chi_A$ for $A = \{x \in X : f(x) = 1\}$. □

Theorem 1.115. *If a set X has n elements and a set Y has p elements, then the set Y^X of functions from X to Y has p^n elements.*

Proof. A function $f : X \to Y$ is determined by the values attributed to each $x \in X$. To see how many functions from X to Y we can form, first note that we can enumerate the elements of X as $x_1, x_2, \ldots x_n$. We have p choices of value for $f(x_1)$, we have p choices of value for $f(x_2)$, and so on. Since these choices are independent, we have p^n choices to attribute images to all elements $x_1, x_2, \ldots x_n$ of X. □

Corollary 1.116. *If a set X has n elements, then its powerset $\mathbb{P}X$ has 2^n elements.*

Proof. In view of Proposition 1.114 and Remark 1.103, $\mathbb{P}X$ and $\{0,1\}^X$ have the same number of elements. By Theorem 1.115, $\{0,1\}^X$ has 2^n elements. □

One may refine the question and ask about the number of subsets with k elements of a given set with n elements.

Definition 1.117. If n and k are non-negative integers, we denote by $\binom{n}{k}$ the number of subsets with k elements of a given set with n elements. We read $\binom{n}{k}$ as "n choose k."

Coefficients of the form $\binom{n}{k}$ are also called *binomial coefficients* for a reason that will become clear with Theorem 1.120 below.

Of course, by definition,

$$\binom{n}{0} = \binom{n}{n} = 1 \tag{1.24}$$

for there is only one subset with 0 element, namely the empty set, and only one subset with n elements, namely the whole set. Similarly,

$$k > n \Longrightarrow \binom{n}{k} = 0.$$

Theorem 1.118. *Let $n \geq k \geq 0$ be integers. The number $\binom{n}{k}$ of subsets with k elements of a given set with n elements is*

$$\binom{n}{k} = \frac{n!}{(n-k)!k!}.$$

Proof. In view of Proposition 1.112 (2), there are

$$n(n-1)\ldots(n-k+1) = \frac{n!}{(n-k)!} \tag{1.25}$$

one-to-one maps from $\{1,\ldots,k\}$ to our given set of cardinality n. The range of each such map is a subset with k elements, and there are $k!$ one-to-one maps with a given subset with k elements as range, by Proposition 1.112 (1). In other words, in (1.25), each subset with k elements is counted $k!$ times. Hence there are

$$\frac{n!}{(n-k)!k!}$$

such subsets. □

Note that for every integers n and k with $n \geq k \geq 0$, we have

$$\binom{n}{k} = \binom{n}{n-k} \tag{1.26}$$

because if $|X| = n$, there are as many ways to pick subsets with k elements than to pick their complements, which are exactly the subsets with $n-k$ elements. Moreover,

$$\sum_{k=0}^{n} \binom{n}{k} = 2^n, \tag{1.27}$$

for there are 2^n subsets of X in all by Corollary 1.116, and each such subset is accounted for in $\sum_{k=0}^{n} \binom{n}{k}$ depending on its cardinality $k \in \{0, 1, \ldots n\}$.

Additionally:

Proposition 1.119. *Let $n \geq k > 0$ be integers. Then*

$$\binom{n}{k} = \binom{n-1}{k-1} + \binom{n-1}{k}.$$

Proof. Let x_0 be a particular point of X, where $|X| = n$. A subset of X with k elements either contains x_0 or does not. If it does, x_0 is already picked, so the number of ways to pick such a set is to pick $k-1$ points out of the remaining $n-1$ points. Hence there are $\binom{n-1}{k-1}$ subsets with k elements, one of which is x_0. On the other hand, a set with k elements none of which is x_0 is obtained by picking k points out of the remaining $n-1$. Hence there are $\binom{n-1}{k}$ of them. Therefore there are

$$\binom{n-1}{k-1} + \binom{n-1}{k}$$

subsets with k elements. □

Proposition 1.119 gives a simple way to find quickly the coefficients $\binom{n}{k}$ for the first few values of n, by building *Pascal's triangle*, in which a row is indexed by n and a column by k where $0 \leq k \leq n$. Proposition 1.119 states that the coefficient in row n and column k is obtained by adding the coefficient in the same column and in the previous column both taken in the preceding row. For $n = 1$, we have $\binom{1}{0} = \binom{1}{1} = 1$ by (1.26), which also gives that each row will start and end with a 1. With this in mind, each row is easily built from the previous one using Proposition 1.119:

$$
\begin{array}{ll}
1\ 1 & (n=1)\\
1\ 2\ 1 & (n=2)\\
1\ 3\ 3\ 1 & (n=3)\\
1\ 4\ 6\ 4\ 1 & (n=4)\\
1\ 5\ 10\ 10\ 5\ 1 & (n=5)\\
1\ 6\ 15\ 20\ 15\ 6\ 1 & (n=6)\\
1\ 7\ 21\ 35\ 35\ 21\ 7\ 1 & (n=7)\\
1\ 8\ 28\ 56\ 70\ 56\ 28\ 8\ 1 & (n=8)\\
\vdots & \vdots
\end{array}
$$

Hence, for instance, $\binom{7}{2}$ is readily seen to be $6+15=21$.

It turns out that $\binom{n}{k}$ is exactly the coefficient of $x^{n-k}y^k$ in the expansion of the binomial $(x+y)^n$, hence the name *binomial coefficient.*

Theorem 1.120 (Binomial Theorem). *The coefficient of $x^{n-k}y^k$ in $(x+y)^n$ is $\binom{n}{k}$:*

$$(x+y)^n = \sum_{k=0}^{n}\binom{n}{k}x^{n-k}y^k = \sum_{k=0}^{n}\binom{n}{k}x^ny^{n-k}.$$

Proof. In multiplying

$$(x+y)^n = (x+y)\cdot(x+y)\cdot\ldots\cdot(x+y),$$

there is $\binom{n}{1}$ ways to pick y in only one factor (and hence x in all $n-1$ other factors), $\binom{n}{2}$ ways to pick y twice among n factors (and thus to pick x in $n-2$ factors), $\binom{n}{3}$ ways to pick y in 3 factors, and so on. In particular, there are $\binom{n}{k}$ ways to pick y in k factors, hence x in $n-k$ factors, that is, $\binom{n}{k}$ is the coefficient of y^kx^{n-k}. □

For instance, looking at the row $n=8$ in Pascal's triangle above, we may apply Theorem 1.120 to the effect that

$$(x+y)^8 = x^8+8x^7y+28x^6y^2+56x^5y^3+70x^4y^4+56x^3y^5+28x^2y^6+8xy^7+y^8.$$

Additional Exercises (Homework)

Exercise 1.121. Are the following one-to-one, onto, both, or neither. Justify your answers:

1. $f:\{a,b,c,d\}\to\{1,2,3,4\}$ defined by $f(a)=1$; $f(b)=4$; $f(c)=3$, and $f(d)=2$.
2. $f:\{a,b,c,d\}\to\{1,2,3,4\}$ defined by $f(a)=1$; $f(b)=4$; $f(c)=3$, and $f(d)=1$.
3. $f:[0,\infty)\to\mathbb{R}$ defined by $f(x)=\sqrt{x}$.
4. $f:[0,\infty)\to[0,\infty)$ defined by $f(x)=\sqrt{x}$.
5. $f:\mathbb{R}\to\mathbb{R}$ defined by $f(x)=\arctan x$.

6. $f : \mathbb{R} \setminus \{0\} \to \mathbb{R}$ defined by $f(x) = \frac{1}{x}$.
7. $f : \mathbb{R} \to \mathbb{R}$ defined by $f(x) = e^x$.
8. $f : \mathbb{R} \to (0,\infty)$ defined by $f(x) = e^x$.
9. $f : \mathbb{R} \to \mathbb{R}$ defined by $f(x) = \sin x$.
10. $f : \mathbb{R} \to [-1,1]$ defined by $f(x) = \sin x$.
11. $f : \left[-\frac{\pi}{2}, \frac{\pi}{2}\right] \to [-1,1]$ defined by $f(x) = \sin x$.
12. $f : \mathbb{N} \to \mathbb{N}$ defined by $f(n) = n^2$.

Exercise 1.122. Let P be a set of 5000 people of weight between 100 and 349 pounds, and consider the function $f : P \to \mathbb{N}$ that associates with $p \in P$ his or her weight $f(p)$ in pounds rounded to the nearest integer.

1. Is f one-to-one, onto, both, or neither?
2. At least how many people in P weigh 200 pounds?

Exercise 1.123. A million pine trees grow in a forest. It is known that no pine tree has more than 600,000 needles. Prove that there are two pine trees in the forest with the same number of needles.

Exercise 1.124. Twenty-five crates of apples are delivered to a store. Each crate contains the same kind of apples, and there are three possible varieties. Prove that among these crates there are at least nine containing the same variety of apples.

Exercise 1.125. How many ways are there of giving out 6 pieces of candy to 3 children (we do not assume that every child receives a piece of candy)?

Exercise 1.126. In Exercise 1.68, how many subsets does $\mathscr{A}$ have?

Exercise 1.127. How many elements does $\mathbb{P}(\mathbb{P}(\mathbb{P}(A)))$ have if A has n elements?

Exercise 1.128. Show that for every integer $n \geq 2$,

$$\binom{n}{2} + \binom{n+1}{2} = n^2.$$

Exercise 1.129. Show that

$$\sum_{k=0}^{n} (-1)^k \binom{n}{k} = 0.$$

1.8 Product Set and Functions

Definition 1.130. The product $X \times Y$ of sets X and Y is the set of ordered pairs ([16]), where the first coordinate is an element of X and the second an element of Y:

$$X \times Y = \{(x,y) : x \in X \wedge y \in Y\}.$$

[16] To formally define ordered pairs within set theory, one may define ordered pairs (i.e., elements of $A \times B$) by

$$(a,b) = \{\{a\},\{a,b\}\},$$

It is often useful to mentally think of $X \times Y$ as in this picture, in which a point (x,y) of $X \times Y$ has a first coordinate or projection $x \in X$ and a second coordinate (or projection) $y \in Y$.

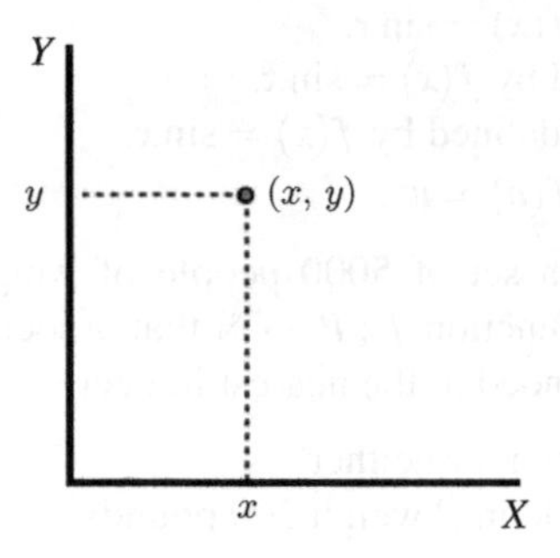

Remark 1.131. Note that formally $X \times Y \neq Y \times X$. For instance, if $X = \{1,2,3\}$ and $Y = \{2,4,5\}$, then $(1,2) \in X \times Y$ but $(1,2) \notin Y \times X$. However, there is a bijection $f : X \times Y \to Y \times X$ defined by $f(x,y) = (y,x)$.

Theorem 1.132. *Let A, B, C, and D be sets. Then*

1. $A \times (B \cup C) = (A \times B) \cup (A \times C)$;
2. $A \times (B \cap C) = (A \times B) \cap (A \times C)$;
3. $A \times \emptyset = \emptyset \times A = \emptyset$;
4. $(A \times B) \cap (C \times D) = (A \cap C) \times (B \cap D)$;
5. $(A \times B) \cup (C \times D) = (A \cup C) \times (B \cup D)$.

Exercise 1.133. Prove Theorem 1.132.

Recall that a *function* $f : X \to Y$ associates with each $x \in X$ exactly one value $y = f(x) \in Y$.

Definition 1.134. The *graph of a function* $f : X \to Y$ is the subset

$$\tilde{f} = \{(x,y) \in X \times Y : y = f(x)\}$$

of $X \times Y$.

Remark 1.135. Since the function f has a uniquely determined graph, and the graph $\tilde{f}$ uniquely determines f, we often identify the function f and its graph $\tilde{f}$. In fact, this is the way to properly define a function as a set, namely, as a subset A of $X \times Y$

---

because this yields the desired property that

$$(a,b) = (c,d) \iff a = c \text{ and } b = d,$$

even though this requires a proof. Namely, if $(a,b) = (c,d)$ we distinguish the case $a = b$ and the case $a \neq b$. In the first case, $\{\{a\},\{a,b\}\} = \{\{a\},\{a\}\} = \{\{a\}\}$ is a singleton and thus so is $\{\{c\},\{c,d\}\}$ so that $c = d = a = b$. In the second case, $(a,b) = \{\{a\},\{a,b\}\}$ is composed of a singleton and a doubleton. If it is equal to $\{\{c\},\{c,d\}\}$, then the singletons must be equal and thus $\{a\} = \{c\}$ and $\{a,b\} = \{c,d\}$ so that $a = c$ and $b = d$.

with the property that for every $x \in X$ there is exactly one $y \in Y$ with $(x,y) \in A$. We write $f(x) = y$ if $(x,y) \in A$, and in this way A determines a unique function f. In fact, formally, the set A *is* the function, even though we prefer in practice to think of f as in Definition 1.98. We do this to have every object ultimately defined within set-theory, like we did for integers in (1.16) and for ordered pairs in Footnote 16 on page 43.

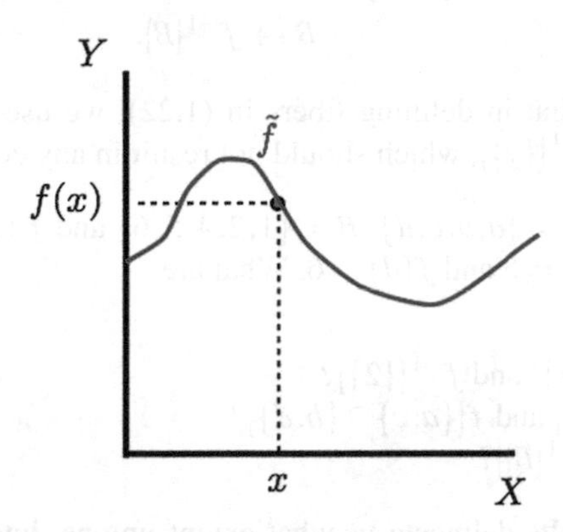

We will say more on this point of view in Chapter 3, when introducing functional relations (Definition 3.21).

Definition 1.136. Given a function $f : X \to Y$ a subset A of X and a subset B of Y, we define the *image* $f[A]$ *of* A *under* f as

$$f[A] = \{y \in Y : \exists a \in A\ f(a) = y\} = \{f(a) : a \in A\}. \tag{1.28}$$

Similarly, the *preimage* $f^{-1}[B]$ *of* B *under* f is

$$f^{-1}[B] = \{x \in X : f(x) \in B\}. \tag{1.29}$$

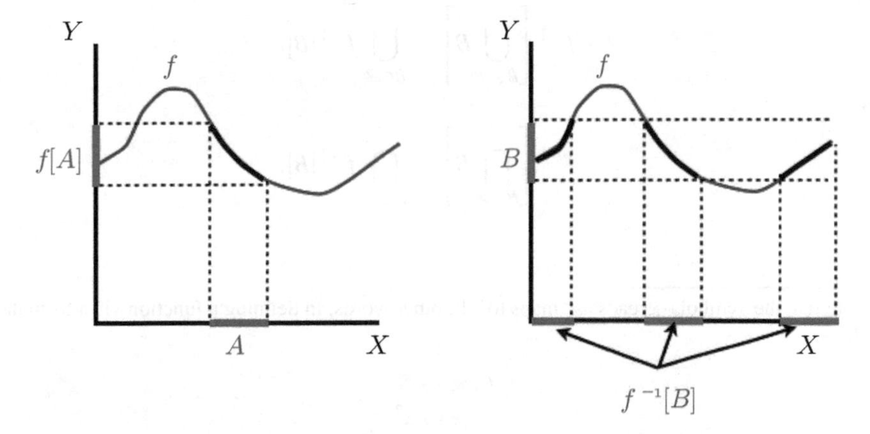

In other words, the function $f : X \to Y$ induces two functions at the level of powersets:

$$\begin{aligned} f[\cdot] : \mathbb{P}X &\to \mathbb{P}Y \\ A &\mapsto f[A] \end{aligned}$$

and ([17])

$$\begin{aligned} f^{-1}[\cdot] : \mathbb{P}Y &\to \mathbb{P}X \\ B &\mapsto f^{-1}[B]. \end{aligned}$$

Remark 1.137. Note that in defining fibers in (1.22), we used the slight abuse of notation $f^{-1}[y]$ for $f^{-1}[\{y\}]$, which should not result in any confusion.

Exercise 1.138. Let $A = \{a,b,c,d\}$, $B = \{1,2,4,5,6\}$ and $f : A \to B$ be defined by $f(a) = f(b) = 1$, $f(c) = 5$ and $f(d) = 6$. What are

1. $f[\{a,b,c\}]$?
2. $f^{-1}[\{1,2\}]$, $f^{-1}[\{1\}]$ and $f^{-1}[\{2\}]$?
3. $f[\{a,c\}] \cap f[\{b,d\}]$ and $f[\{a,c\} \cap \{b,d\}]$?
4. $f^{-1}[f[A]]$ and $f[f^{-1}[B]]$?

The following results delineate to what extent unions, intersections, and complements are preserved or not under direct image or preimage. We prove here (1) through (5), which are nothing but straightforward rephrasing of the definitions. The proofs of (6) through (11) are postponed to the next chapter (see Exercise 2.32), after we discuss proof techniques in more detail.

Theorem 1.139. *If $f : X \to Y$, $A \subset X$, $B \subset Y$, $\mathscr{A} \subset \mathbb{P}X$, and $\mathscr{B} \subset \mathbb{P}Y$, then*

1.

$$f\left[\bigcup_{A \in \mathscr{A}} A\right] = \bigcup_{A \in \mathscr{A}} f[A].$$

2.

$$f^{-1}\left[\bigcup_{B \in \mathscr{B}} B\right] = \bigcup_{B \in \mathscr{B}} f^{-1}[B].$$

3.

$$f^{-1}\left[\bigcap_{B \in \mathscr{B}} B\right] = \bigcap_{B \in \mathscr{B}} f^{-1}[B].$$

---

[17] Note that the symbol $\mapsto$ reads as "maps to." In other words, in defining a function via a formula such as

$$\begin{aligned} f : \mathbb{R} &\to \mathbb{R} \\ x &\mapsto x^2 \end{aligned}$$

the first line specifies the domain and codomain of the function, while the second states that f maps x to x^2, that is, states that $f(x) = x^2$.

4. In general, we may have

$$f[A_1 \cap A_2] \neq f[A_1] \cap f[A_2].$$

5.

$$f^{-1}[Y \setminus B] = X \setminus f^{-1}[B]$$

6. f is one-to-one if and only if $f[X \setminus A] \subset Y \setminus f[A]$ for every $A \subset X$.
7. f is onto if and only if $Y \setminus f[A] \subset f[X \setminus A]$ for every $A \subset X$.
8.

$$f[f^{-1}[B]] \subset B.$$

9.

$$\left(\forall B \subset Y \;\; f[f^{-1}[B]] = B\right) \iff f \text{ is onto.}$$

10.

$$A \subset f^{-1}[f[A]].$$

11.

$$\left(\forall A \subset X \;\; A = f^{-1}[f[A]]\right) \iff f \text{ is one-to-one.}$$

Note that by (6) and (7), f does not preserve complements in general, except if f is a bijection, in contrast to f^{-1}.

Proof. (1) $y \in f[\bigcup_{A\in\mathscr{A}} A]$ if and only if there is $x \in \bigcup_{A\in\mathscr{A}} A$ with $f(x) = y$, if and only if there is $A \in \mathscr{A}$ and $x \in A$ with $f(x) = y$. Hence $y \in f[\bigcup_{A\in\mathscr{A}} A]$ if and only if $y \in \bigcup_{A\in\mathscr{A}} f[A]$.

(2) $x \in f^{-1}[\bigcup_{B\in\mathscr{B}} B]$ if and only if $f(x) \in \bigcup_{B\in\mathscr{B}} B$, that is, if and only if there is $B \in \mathscr{B}$ with $f(x) \in B$, that is, if and only if $x \in \bigcup_{B\in\mathscr{B}} f^{-1}[B]$.

(3) $x \in f^{-1}[\bigcap_{B\in\mathscr{B}} B]$ if and only if $f(x) \in \bigcap_{B\in\mathscr{B}} B$, that is, if $f(x) \in B$ for every $B \in \mathscr{B}$, equivalently, if $x \in \bigcap_{B\in\mathscr{B}} f^{-1}[B]$.

(4) Take, for instance, $X = Y = \{1,2\}$ and $f : X \to Y$ with $f(1) = f(2) = 1$. Let $A_1 = \{1\}$ and $A_2 = \{2\}$. Then $A_1 \cap A_2 = \emptyset$ so that $f[A_1 \cap A_2] = \emptyset$ but $f[A_1] = f[A_2] = f[A_1] \cap f[A_2] = \{1\}$.

(5) $x \in f^{-1}[Y \setminus B]$ if and only if $f(x) \in Y \setminus B$, that is, $f(x) \notin B$. Moreover, $f(x) \notin B$ if and only if $x \notin f^{-1}[B]$, if and only if $x \in X \setminus f^{-1}[B]$. □

Remark 1.140. We use square brackets to lift the ambiguity between the image of a point of X and the image of a subset of X under a function $f : X \to Y$. However, context is normally enough to avoid ambiguity and you will often see the notations $f(A)$ and $f^{-1}(B)$ for what we have denoted $f[A]$ and $f^{-1}[B]$.

Additional Exercises (Homework)

Exercise 1.141. Write explicitly $X \times Y$ and $Y \times X$ in the following cases:
1. $X = \{2,5,6\}$ and $Y = \{c,d\}$.
2. $X = \{\{1,2\},\{3,4\}\}$ and $Y = \{\{a\},\{2,3\},\{a,4\}\}$.

Exercise 1.142. For the sets A, B, C, and D of Exercise 1.94, find:

1. $(D\setminus C)\times(B\setminus A)$.
2. $\mathbb{P}(B)\times\mathbb{P}(C\cap D)$.

Exercise 1.143. Let $G=\{(x,\sin x):x\in\mathbb{R}\}$ denote the graph of the function $\sin:\mathbb{R}\to\mathbb{R}$. Of course, $G\subset\mathbb{R}\times\mathbb{R}$. Sketch the following subsets of $\mathbb{R}\times\mathbb{R}=\mathbb{R}^2$:

1. $G\cap\{(x,y)\in\mathbb{R}^2:x\geq 0\}$.
2. $G\cap\{(x,y)\in\mathbb{R}^2:y<x\}$.
3. $G\cap\{(x,y)\in\mathbb{R}^2:y=1\}$.
4. $\mathbb{R}^2\setminus G$.
5. $([0,1]\times[0,1])\triangle G$.
6. $G\cap\{(x,y)\in\mathbb{R}^2:x^2-x=0\}$.

Exercise 1.144. Let $X=\{1,2,3,4\}$ and $Y=\{\triangle,\bigcirc,\square,\diamondsuit\}$. Let $f:X\to Y$ be defined by $f(1)=f(2)=\square$, $f(3)=\triangle$ and $f(4)=\bigcirc$. Write explicitly

1. The list of points of the graph of f.
2. $f[\{1,3\}]$ and $f^{-1}[f[\{1,3\}]]$.
3. $f^{-1}[\{\diamondsuit,\square\}]$ and $f[f^{-1}[\{\diamondsuit,\square\}]]$.

Suggested Further Readings

Similar material with more emphasis on functions is covered in [13, Chapter 1]. It is also a source of additional exercises.

[3] is an interesting introduction to *elementary set theory* from the historical perspective. At this point of our development of the material, this is an appropriate supplementary reading ([18]) up to Section 3.6 (not included). The latter part of this document treats aspects of our Chapters 3 and 4. The document includes a number of elementary exercises that you should work out along the way. It develops the material in a naive way, as it was originally developed, and then points out the limitations of this naive approach by presenting Russell's paradox (Section 2.7).

The other side of the coin developed in our first chapter is that of *elementary symbolic logic*. This too is developed from an historical perspective in [2] in a similar fashion ([19]) to [3] for set theory. [2] is a good source of additional exercises at the appropriate level on Sections 1.2 and 1.3.

The book [7] introduces Set Theory in a more complete and systematic fashion, but remains at an introductory level. It could be an adequate complement at the end of this course. Other Set Theory books provide a much more complete and advanced

[18]There are a few choices of notations in this document that are not consistent with ours. For instance, like a number of other references [3] uses $\subseteq$ for the subset relation and reserves $\subset$ for proper subsets.

[19]Including discrepancies of notations. For instance, this reference uses $\sim$ for the negation, where we use $\neg$.

treatment of set theory and typically go far beyond what we will cover in this course. However, the first chapter of [8] provides some additional discussion of the need for an axiomatic approach that can shed more light on the issue. The chapter also covers the basic language of set theory, albeit in a quicker and more formal fashion. Chapter 2 of [8] treats essentially the material of our Chapter 3 and can be a good source to review the same material when written for a more mathematically mature audience.

[22] is another set theory book written with undergraduate students in mind. This book too covers quite a bit more than we will, and jumps to our Chapter 4 very quickly.

Chapter 2
On Proofs and Writing Mathematics

Most advanced mathematics books are structured around *definitions*, *theorems*, *propositions*, *lemmas*, *corollaries*, and *proofs*. As discussed before, a definition gives an unambiguous meaning to a mathematical term or symbol. Theorems, propositions, and lemmas are true statements relating the notions defined via definitions. Theorems are statements of significance. Propositions are of lesser significance, but of interest on their own. A lemma is a statement that is often a technical step towards proving a theorem, and is often of questionable interest on its own. A *corollary* is an easy consequence of the previous statement.

Proofs are clear, convincing arguments showing the unquestionable validity of the statements (theorems, propositions, and lemmas). We have already seen proofs in the previous chapter, without much commenting. As a result, you may have found some of these proofs difficult to follow.

In this chapter, we will talk a lot more about proof structures, proof strategies, and "style" in writing proofs. Of course, there is no talking about proofs in vacuum, so we will be going through commented proofs. The context will be in part the set-theoretic notions we have studied thus far, in part other familiar mathematical notions, particularly in arithmetic, that give us a natural framework for relatively easy proofs.

To this end, recall that $\mathbb{N}, \mathbb{Z}, \mathbb{Q}$, and $\mathbb{R}$ denote, respectively, the sets of natural numbers, integers, rational numbers, and real numbers.

Definition 2.1. An integer k is *even* if ([1]) there is $n \in \mathbb{Z}$ with $k = 2n$. An integer is *odd* if it is not even.

---

[1]Note that in the context of a definition "if" really means "exactly if" or "if and only if."

F. Mynard, *An Introduction to the Language of Mathematics*,
https://doi.org/10.1007/978-3-030-00641-9_2

Since each integer n has a successor $n+1$, it is clear that between two consecutive even numbers $2n$ and $2n+2$, there is an odd number $2n+1$. In fact, it is a reformulation of the definition to observe that

$$k \text{ is odd} \iff \exists n \in \mathbb{Z}\,(k = 2n+1)\,. \tag{2.1}$$

Definition 2.2. Let a and b be integers. We say that a *divides* b, in symbols $a|b$, if there is an integer c with $b = ac$. We also say that b is *a multiple of* a and that a is a *divisor* of b.

A natural number $n \geq 2$ is *prime* if its only positive divisors are 1 and n, and is called *composite* if it is not prime.

Many mathematical statements are formally conditional statements (or biconditional statements that can be seen as the conjunction of two conditional statements). Recall that $p \Longrightarrow q$ is automatically true if p is false. Hence, what needs to be verified to check the validity of a conditional statement $p \Longrightarrow q$ is that q is true under the assumption that p is true.

There are three basic strategies to prove $p \Longrightarrow q$:

1. A *direct proof*: Assume p and show q.
2. A *contrapositive proof*: We have seen that $p \Longrightarrow q$ is logically equivalent to its contrapositive $\neg q \Longrightarrow \neg p$. So we may assume the negation of q and show the negation of p. This may seem convoluted but it is sometimes much easier to show than the original conditional statement.
3. A *proof by contradiction*: Assume that p is true and show that if q were false, we could obtain as a logical consequence a contradictory statement. That is enough to conclude that the assumption that q is false is wrong, hence that q is true.

2.1 Direct Proofs

The general pattern here is to prove a statement of the form

If P, then Q

in a direct fashion, that is, showing Q under the assumption that P is true. Hence the proof template would look like

Proof. Suppose P.

$\vdots$

Therefore Q. □

Here we only need to fill the gap symbolized by $\vdots$. The word *Proof* indicates the beginning of the proof, while the symbol □ indicates its end.

Let us now study some simple examples:

Proposition 2.3. *If x is odd, then x^2 is odd.* ([2])

Proof. Suppose that x is odd ([3]). In view of (2.1), there is $n \in \mathbb{Z}$ with $x = 2n+1$. ([4]) Thus

$$x^2 = (2n+1)^2 = 4n^2 + 4n + 1 = 2(2n^2 + 2n) + 1.$$

([5]) Letting $k = 2n^2 + 2n$, we have $x^2 = 2k+1$ for $k \in \mathbb{Z}$. Therefore x^2 is odd ([6]). □

Exercise 2.4. Prove that if x is odd then $x+1$ is even.

Let us examine a few more examples.

Proposition 2.5. *Let a, b, and c be integers. If $a|b$ and $b|c$, then $a|c$.* ([7])

Proof. Suppose that $a|b$ and $b|c$. ([8]) By definition, there are integers k and n with $b = ak$ and $c = bn$. Thus ([9])

$$c = bn = (ak)n = a(kn),$$

so that $c = am$ for some integer $m = kn$. Therefore $a|c$. □

Proposition 2.6. *Let a,b, and n be integers. If $n|a$ and $n|b$, then $n|(a-b)$.*

Exercise 2.7. Prove Proposition 2.6.

Sometimes, in the course of a proof (whether direct or not), we need to distinguish all the possible cases or alternatives, and sometimes we need to consider these alternatives repeatedly.

---

[2] Often, when a variable such as x appears, it is good to specify in the statement what kind of object x is; here, it is an integer, but it is somewhat redundant to state it, for *odd* is a term that only applies to integers. In other words, when writing mathematics, you want to say enough to rule out ambiguities, but avoid saying more than needed.

[3] This is part of the standard setup for a direct proof: the "Suppose P." of the template.

[4] Here we simply write out what the assumption that x is odd means. This is often how you start out: by simply spelling out definitions. Next we try to fill in the gap to conclude that x^2 is odd. We have written something about the form of x, so we try to derive something about the form of x^2. The natural thing to do is to square both sides.

[5] While the second equality sign is simply foiling the square out, the third stems from what we are looking for. We are trying to identify the pattern $2\square+1$ for the form of x^2, so we factor 2 in the two first terms.

[6] That conclusion is also part of our template and can be written ahead of time to make sure to keep in mind "where you want to go."

[7] In other words, in the terminology of Chapter 3, the relation $|$ is transitive.

[8] This is again part of the standard setup of a direct proof, which you should complement with writing out your conclusion after leaving some space, to know where you are going. Like in the previous example, we will start by spelling out what our assumption means.

[9] This is the only step where we need to do something: combine the two pieces of information rephrased from the assumption, and try to match the desired form for the conclusion that $a|c$, that is, that there is $m \in \mathbb{Z}$ with $c = am$.

Theorem 2.8. *If n is a natural number greater than* 1*, then n has at least one prime divisor.*

Proof. Let $n > 1$ be a natural number. If n is prime, we are done, because n divides n. If not ([10]), n has a divisor a_1 greater than 1 but less than n. The same reasoning applies to a_1 (where $1 < a_1 < n$): either a_1 is prime and we are done, or it is not and it has a divisor a_2 with $1 < a_2 < a_1 < n$. The sequence $\{a_1, a_2, \ldots\}$ needs to terminate with a prime number in finitely many steps, for there are only finitely many natural numbers between 1 and n. Hence n has a prime divisor. □

Now we turn to an example formulated in set-theoretic terms.

Example 2.9. Prove that

$$\{x \in \mathbb{Z} : 15|x\} \subset \{x \in \mathbb{Z} : 3|x\}. \tag{2.2}$$

Here we want to prove a set-theoretic inclusion, but as we have repeatedly said, this is simply an instance of conditional statement, namely that

$$15|x \Longrightarrow 3|x.$$

Yet formally, you might want to start out by taking an element of the set to the left in (2.2), that is, take an $x \in \mathbb{Z}$ with $15|x$, and try to show that $3|x$.

Proof. Let $t \in \{x \in \mathbb{Z} : 15|x\}$, that is, let $t \in \mathbb{Z}$ with $15|t$. ([11]) In other words ([12]), there is an integer k with

$$t = 15k = 3 \times 5 \times k = 3(5k),$$

([13]) that is, $t = 3m$ for the integer $m = 5k$. Therefore $3|t$, that is, $t \in \{x \in \mathbb{Z} : 3|x\}$. Thus

$$\{x \in \mathbb{Z} : 15|x\} \subset \{x \in \mathbb{Z} : 3|x\}.$$

□

Example 2.10. Prove that

$$\{x \in \mathbb{Z} : 2|x\} \cap \{x \in \mathbb{Z} : 9|x\} \subset \{x \in \mathbb{Z} : 6|x\}.$$

Proof. Let $t \in \{x \in \mathbb{Z} : 2|x\} \cap \{x \in \mathbb{Z} : 9|x\}$, that is, $2|t$ and $9|t$. This means that there are integers k and n with

$$t = 2k = 9n.$$

[10] Here we look at the two possible alternatives.

[11] Note that we picked t rather than x, because x is the generic variable used to describe elements of the set. As we pick a particular element, it is better, though not necessary, to use a different symbol to distinguish this element.

[12] As usual, we start with spelling out what the assumption means according to our definitions.

[13] The reason to factor out 3 is clear if you keep in mind what we want to establish: that x can be written under the form $x = 3m$ for some integer m.

Since t is even, $9n$ is even and thus n is also even (otherwise $9n$ would be odd). Thus $n = 2p$ for some integer p. Therefore

$$t = 9n = 9 \times 2p = 18p = 6(3p),$$

so that $t = 6r$ for the integer $r = 3p$. Thus $6|t$, that is, $t \in \{x \in \mathbb{Z} : 6|x\}$. Therefore

$$\{x \in \mathbb{Z} : 2|x\} \cap \{x \in \mathbb{Z} : 9|x\} \subset \{x \in \mathbb{Z} : 6|x\}.$$

□

Proposition 2.11. *If x and y are positive real numbers, then*

$$2\sqrt{xy} \leq x + y. \tag{2.3}$$

Let us start with some preparatory work/thinking before writing a formal proof. We might want to try to restate (2.3) in a different form. For instance, squaring both sides of (2.3) yields

$$4xy \leq (x+y)^2 = x^2 + 2xy + y^2, \tag{2.4}$$

which, subtracting $4xy$ on each side, is equivalent to

$$0 \leq x^2 - 2xy + y^2.$$

By observing that $x^2 - 2xy + y^2$ is nothing but $(x-y)^2$, we note that (2.4)—obtained from (2.3) by squaring—is equivalent to the transparently true statement $(x-y)^2 \geq 0$. Of course, this does not guarantee yet that (2.3) is true, because we only had

$$(2.3) \Longrightarrow (2.4)$$

by squaring. We would obtain the converse (2.4) $\Longrightarrow$ (2.3) if taking the square root on both sides of (2.4) would preserve the order, resulting in (2.3). That we even *can* take the square root of both sides is guaranteed by the assumption that both x and y are positive, so that $4xy$ is positive too. These considerations give us a strategy: prove as an intermediate step that $f(x) = \sqrt{x}$ is an order-preserving function, that is, a non-decreasing function, then start from the true statement $(x-y)^2 \geq 0$ and derive (2.3):

Lemma 2.12. *If x and y are positive real numbers and $x \leq y$, then $\sqrt{x} \leq \sqrt{y}$.* ([14])

[14] This could be proved using results from Calculus, namely that a differentiable function is increasing on an interval on which it has a positive derivative. Since the derivative of $f(x) = \sqrt{x} = x^{\frac{1}{2}}$ is

$$f'(x) = \frac{1}{2}x^{-\frac{1}{2}} = \frac{1}{2\sqrt{x}} \geq 0$$

for all $x > 0$, f is increasing on $(0, \infty)$ and the result follows. But we want to adopt a more pedestrian yet slightly trickier approach, trying to prove everything essentially from scratch.

Proof. Assume $x \leq y$. Then $x - y \leq 0$. Since x and y are positive numbers, they can be interpreted as squares, namely, $x = (\sqrt{x})^2$ and $y = (\sqrt{y})^2$ so that $x - y$ can be thought of as a difference of squares, which factors as

$$x - y = (\sqrt{x})^2 - (\sqrt{y})^2 = (\sqrt{x} - \sqrt{y})\,(\sqrt{x} + \sqrt{y})\,.$$

Thus $(\sqrt{x} - \sqrt{y})\,(\sqrt{x} + \sqrt{y})$ is negative or zero, and $(\sqrt{x} + \sqrt{y}) > 0$, so that $\sqrt{x} - \sqrt{y} \leq 0$, that is,

$$\sqrt{x} \leq \sqrt{y}.$$

□

We called this result a lemma, because we stated it and proved it for the sole purpose of proving Proposition 2.11. Hence in this context, it is a lemma.

Proof (Proof of Proposition 2.11). Suppose x and y are positive real numbers. Then $(x - y)^2 \geq 0$, that is,

$$x^2 - 2xy + y^2 \geq 0.$$

Adding $4xy$ on both sides ([15]), we obtain

$$x^2 + 2xy + y^2 = (x + y)^2 \geq 4xy.$$

Since x and y are positive, both sides of the inequality are positive, so that Lemma 2.12 applies to the effect that

$$\sqrt{(x + y)^2} \geq \sqrt{4xy} = 2\sqrt{xy}.$$

As $\sqrt{(x + y)^2} = |x + y| = x + y$, we conclude that

$$2\sqrt{xy} \leq x + y.$$

□

Exercise 2.13. Prove Proposition 1.102.

A Word on Biconditional Statements

Before turning to the next section, let me reiterate that to prove a biconditional statement (equivalence), we can prove two conditional statements, as seen in (1.7):

$$(p \iff q) \equiv (p \Longrightarrow q) \wedge (q \Longrightarrow p)\,.$$

---

[15]This step could look like it is coming out of the blue without the preparatory work, but we are simply retracing our steps from the preparatory work.

Hence we may show an equivalence in a direct fashion, by showing two conditional statements in a direct way, as discussed so far. To illustrate this, let us return to set-theoretic statements, as we did not go over many such examples. Recall also that equality of sets is often proved by showing two different inclusions, as pointed out in (1.18):

$$X = Y \iff X \subset Y \wedge Y \subset X.$$

Example 2.14. Show that

$$B \setminus (B \setminus A) = A \iff A \subset B.$$

Solution. We first show the direction $\Longrightarrow$. To this end, assume $B \setminus (B \setminus A) = A$. To show $A \subset B$, let $x \in A$ ([16]). Since

$$x \in A = B \setminus (B \setminus A) \text{ and } B \setminus (B \setminus A) \subset B,$$

we conclude that $x \in B$. Hence $A \subset B$.

We now prove the converse (the direction $\Longleftarrow$). Assume that $A \subset B$. We will show $B \setminus (B \setminus A) = A$ by double inclusion. To show $A \subset B \setminus (B \setminus A)$, let $x \in A$. Then $x \in B$ because $A \subset B$. Since $x \notin B \setminus A$ because $x \in A$, we conclude that $x \in B \setminus (B \setminus A)$.

To show the reverse inclusion, let $x \in B \setminus (B \setminus A)$, that is, $x \in B$ and $x \notin B \setminus A$. In view of Proposition 1.79,

$$x \in (B \cap A^c)^c = B^c \cup A,$$

so that $x \in A$ because $x \notin B^c$. Thus $B \setminus (B \setminus A) \subset A$.

Exercise 2.15. Show that:

1.
$$B \setminus A = B \iff B \cap A = \emptyset.$$

2.
$$(A \cup B) \cap (A \cup B^c) = A.$$

3.
$$A \subset B \iff A \cap B = A \iff A \cup B = B.$$

Sometimes, a biconditional statement can be proved with a sequence of equivalences, such as the proof of Theorem 1.132 in Exercise 1.133.

---

[16] As $A \subset B$ means that

$$x \in A \Longrightarrow x \in B,$$

we start with $x \in A$ and we have to show that $x \in B$.

Additional Exercises (Homework)

Exercise 2.16. Show that if n is an odd integer, then n^3 is also odd.

Exercise 2.17. Show that if a and b are integers such that $a|b$, then $a^2|b^2$.

Exercise 2.18. Show that if a,b,c are positive integers and $ac|bc$, then $a|b$.

Exercise 2.19. Show that if two integers have opposite parity (that is, one is even the other odd) then their product is even.

Exercise 2.20. Show that $5n^2+3n+7$ is odd for every $n \in \mathbb{Z}$.

Exercise 2.21. Let $x,y \in \mathbb{R}$. Show that if

$$x^2+5y = y^2+5x$$

then either $x = y$ or $x+y = 5$.

Exercise 2.22. Show that if $x \in (0,4) \subset \mathbb{R}$ then

$$\frac{4}{x(4-x)} \geq 1.$$

Exercise 2.23. Prove that if $x^2 \leq 1$ then $x^2-7x > -10$.

Exercise 2.24. Let n be an odd integer. Show that either $n = 4j+1$ for some integer j or $n = 4i-1$ for some integer i.

2.2 Contrapositive Proofs

Since

$$p \Longrightarrow q \equiv \neg q \Longrightarrow \neg p,$$

we may show $\neg q \Longrightarrow \neg p$ to show $p \Longrightarrow q$. This is a *contrapositive proof.* The template is then to assume $\neg q$ and to bridge the gap to reach the conclusion $\neg p$.

Example 2.25. Prove that if n is an integer such that $7n+9$ is even then n is odd.

Proof. ([17]) Assume that n is not odd, that is, that n is even, that is, $n = 2k$ for some integer k. Then

$$7n+9 = 7(2k)+9 = 14k+9 = 2(7k+4)+1$$

is odd for $7n+9 = 2p+1$ for the integer $p = 7k+4$. Thus $7n+9$ is not even. □

[17] We proceed by contrapositive so we assume the negation of the conclusion, that is, that n is not odd, that is, n is even.We want to conclude the negation of "$7n+9$ is even," that is, "$7n+9$ is odd."

Exercise 2.26. Try a direct proof for Example 2.25.

Example 2.27. Prove that if x^2-6x+5 is even then x is odd.
Note that a direct proof would be particularly cumbersome.

Proof. Assume that x is not odd, that is, that x is even. Then $x=2k$ for some integer k. Therefore

$$\begin{aligned} x^2-6x+5 &= (2k)^2-6(2k)+5 = 4k^2-12k+5 \\ &= 2\left(2k^2-6k+2\right)+1 = 2p+1 \end{aligned}$$

for the integer $p=2k^2-6k+2$. Thus x^2-6x+5 is not even. □

Exercise 2.28. Show that if m^2 is even then m is even.

Example 2.29. Let n and p be integers. Show that if $n \geq 2$ then $n \nmid p$ or $n \nmid (p+1)$.

Proof. We prove the contrapositive. Assume that $n \nmid p$ or $n \nmid (p+1)$ is false, that is (using DeMorgan's laws), assume that $n \mid p$ **and** $n \mid p+1$. By Proposition 2.6, $n \mid (p+1-p)$, that is, $n \mid 1$. Therefore there is $k \in \mathbb{Z}$ with $kn=1$ and thus $n=\pm 1$ is not greater or equal to 2. □

Example 2.30. Let x and y be real numbers. Show that if $xy^4+x^3 \geq y^5+x^2y$ then $y \leq x$.

Note that here, like in Example 2.27, it would be hard to know what to do with the assumption that $xy^4+x^3 \geq y^5+x^2y$ if we were to try a direct argument. So we try the contrapositive and start with the simpler assumption that $y \leq x$ is not true, that is, the assumption that $y > x$. We then want to get to the conclusion that

$$xy^4+x^3 < y^5+x^2y.$$

In preparatory work, we sometimes work back from the desired conclusion:

$$\begin{aligned} xy^4+x^3 < y^5+x^2y &\iff y^5+x^2y-xy^4-x^3 > 0 \\ &\iff y(y^4+x^2)-x(y^4+x^2) > 0 \\ &\iff (y-x)(y^4+x^2) > 0, \end{aligned}$$

which gives us a clear handle on how to proceed: from the positive quantity $y-x$ (we assumed $y > x$), we multiply both sides by the positive quantity y^4+x^2 to work our way to the desired conclusion.

Proof. We proceed by contraposition. Assume that $y \leq x$ is false, that is, that $y > x$. Then $y-x > 0$ and $y^4+x^2 > 0$ for x and y are not equal and thus not both zero. Thus the product

$$(y-x)(y^4+x^2) > 0,$$

that is,

$$y^5+x^2y-xy^4-x^3 > 0,$$

equivalently,

$$y^5 + x^2y > xy^4 + x^3,$$

that is, $xy^4 + x^3 \geq y^5 + x^2y$ is false. □

Exercise 2.31. Let x and y be real numbers such that $x < 2y$. Prove that if $7xy \leq 3x^2 + 2y^2$ then $3x \leq y$.

We are now in a better position to revisit Theorem 1.139.

Exercise 2.32. Show (6) through (11) in Theorem 1.139.

Additional Exercises (Homework)

Exercise 2.33. Show that if $n \in \mathbb{Z}$ and n^2 is odd then n is odd.

Exercise 2.34. Show that if a product of two positive real numbers is greater than 144, then at least one of them is greater or equal to 12.

Exercise 2.35. Show that if n and p are integers and $n + p > 21$ then $n \geq 11$ or $p \geq 11$.

Exercise 2.36. Let n, p be integers. Show that if $n^2(p^2 - 2p)$ is odd then n and p are odd.

Exercise 2.37. Let n, p be integers. Show that if np is irrational then n or p is irrational.

Exercise 2.38. Let n and p be integers. Show that :

1. if np is even, then n or p is even.
2. if np is odd, then n and p are odd.

Exercise 2.39. Let $n \in \mathbb{Z}$. If $3 \not| n^2$, then $3 \not| n$.

Exercise 2.40. Let $x \in \mathbb{R}$. Show that if $x^3 - x > 0$ then $x > -1$.

Exercise 2.41. Let $n \in \mathbb{Z}$. Show that if $4 \not| n^2$ then n is odd.

Exercise 2.42. Let $x \in \mathbb{R}$. Show that if

$$x^5 + 7x^3 + 5x \geq x^4 + x^2 + 8$$

then $x \geq 0$.

2.3 Proofs by Contradiction

A proof by contradiction could potentially be used to prove any statement, whether it is formally a conditional statement or not.

The idea is simply to show that if the result was not true, then something clearly false (a contradiction) would also have to be true. Formally, to prove P by contradiction, you assume $\neg P$ and show that it implies a false statement (a contradiction). The canonical form of a contradiction is $Q \wedge \neg Q$. So the pattern is typically to come up (!) with a statement Q for which you can prove

$$\neg P \Longrightarrow Q \wedge \neg Q. \tag{2.5}$$

If you proved the conditional statement (2.5), then P has to be true, for if it wasn't, then $Q \wedge \neg Q$ would be true too, which is absurd. Note that Q may not have anything to do with P at all! So this is a strategy that sometimes requires creativity in coming up with a useful condition Q.

Consider, for instance, the following classical fact:

Theorem 2.43. *The set of prime numbers is infinite.*

This is not a conditional statement. There is no premise to try to use, and the strategies we have discussed in the two previous sections are of no use.

To consider another such example, recall that a real number x is *rational* if there is $p \in \mathbb{Z}$ and $q \in \mathbb{Z}\backslash\{0\}$ with $x = \dfrac{p}{q}$ (see Section 3.4 for a more thorough treatment of rationals). An *irrational number* is a real number that is not rational. If $x = \dfrac{p}{q}$ is rational, then we can cancel all common factors among p and q and obtain x *in lowest terms*.

Proposition 2.44. *The number $\sqrt{2}$ is irrational.*

This is a similarly non-conditional statement, which prevents us from using the approaches of the previous two sections.

Both Theorem 2.43 and Proposition 2.44 are of historical significance and were proved by ancient Greeks, essentially along the lines of the proofs presented here, even though Euclid's argument for Theorem 2.43 was not technically an argument by contradiction (see, for instance, [6]). Greeks were thinking of numbers in terms of their geometric interpretation. Hence $\sqrt{2}$ was a perfectly natural quantity to consider, as the length of the diagonal of a square of side 1. Followers of Pythagoras, after his discovery of the role of proportions—hence fractions—in musical harmony, developed a belief that the universe was ultimately governed by rational quantities. Read, for instance, [5]. Thus the discovery of Proposition 2.44 was a shock that called into question their worldview.

Before we return to these historical proofs of non-conditional statements, let us examine a (simpler) proof by contradiction of a simple conditional statement.

Proposition 2.45. *If p is an integer, then $n^2 - 4p \neq 2$ for every real number n.*

Proof. We proceed by contradiction. Suppose then that p is an integer and n is a real number such that

$$n^2 - 4p = 2, \tag{2.6}$$

equivalently,

$$n^2 = 2 + 4p = 2(2p+1).$$

Since $2p+1$ is an integer, n^2 is an even integer, so that n is also even (see Exercise 2.28), that is, $n = 2k$ for some integer k. Substituting in (2.6) yields

$$4k^2 - 4p = 2,$$

so that

$$2k^2 - 2p = 1 = 2(k^2 - p).$$

As a result, 1 is even, for $k^2 - p$ is an integer. But 1 is odd ([18]). This is a contradiction with our assumption (2.6), which is therefore false. □

Let us also consider a simple example of a statement formulated in set-theoretic terms:

Example 2.46. Show that the graphs $y = x^2 + x + 4$ and $y = x - 3$ do not intersect.

Proof. Assume to the contrary that $y = x^2 + x + 4$ and $y = x - 3$ intersect ([19]), that is, that there is a point (a, b) of the plane with

$$b = a^2 + a + 4 = a - 3,$$

so that $a^2 = -7 < 0$. But $a^2 \geq 0$ ([20]). This is a contradiction with our assumption that the graphs intersect. □

Let us now return to Theorem 2.43 and Proposition 2.44.

Proof (Proof of Proposition 2.44). We proceed by contradiction. Assume that $\sqrt{2}$ is rational. Then there are integers p and $q \neq 0$ with

$$\sqrt{2} = \frac{p}{q},$$

and we can assume without loss of generality that p and q do not have any common factor (we cancel them if they do). By squaring both sides, we obtain

$$2 = \frac{p^2}{q^2}$$

[18] Here the statement Q of (2.5) can be picked to be either "1 is even" or its negation "1 is odd."

[19] That is $\neg P$ in (2.5).

[20] Here Q is the statement that $a^2 \geq 0$ and $\neg Q$ that $a^2 < 0$.

so that $p^2 = 2q^2$ is even. Thus p is also even (see Exercise 2.28), that is, $p = 2k$ for some integer k. Hence

$$p^2 = 4k^2 = 2q^2$$

and thus $q^2 = 2k^2$ is also even. Thus q is also even. It means that p and q have a common factor 2, in contradiction to our assumption that $\sqrt{2} = \dfrac{p}{q}$ is in lowest terms. Thus $\sqrt{2}$ is irrational. □

In the same spirit, defining the *Euler number e* by ([21])

$$e = \sum_{n=0}^{\infty} \frac{1}{n!}, \tag{2.7}$$

we have:

Proposition 2.47. *The Euler number is irrational.*

Note that (2.7) is a convergent series by the Ratio Test, and moreover

$$1 + \frac{1}{1!} = 2 < e < 1 + \sum_{n=0}^{\infty} 2^{-n} = 1 + 2 = 3,$$

because

$$\sum_{n=1}^{\infty} \frac{1}{n!} = 1 + \frac{1}{2} + \frac{1}{6} + \frac{1}{24} \ldots < \sum_{n=0}^{\infty} \frac{1}{2^n} = 1 + \frac{1}{2} + \frac{1}{4} + \frac{1}{8} \ldots$$

Proof. Assume to the contrary that $e = \frac{p}{q}$ (where we can assume $p > q > 1$ for $e > 1$ is not a natural number). Then

$$x = q!\left(e - \sum_{n=0}^{q} \frac{1}{n!}\right)$$

is an integer. Indeed, $q!e = p(q-1)!$ is an integer and so is $q!\sum_{n=0}^{q} \frac{1}{n!}$ as a sum of integers.

Moreover, $x > 0$ for

$$x = q!\left(\sum_{n=0}^{\infty} \frac{1}{n!} - \sum_{n=0}^{q} \frac{1}{n!}\right) = q! \sum_{n=q+1}^{\infty} \frac{1}{n!} > 0.$$

We claim that $x < 1$. To see this, note that for each $n > q$,

$$\frac{q!}{n!} = \frac{1}{(q+1)(q+2)\ldots(n-1)n} = \frac{1}{\prod_{i=1}^{n-q}(q+i)},$$

[21] We are of course assuming familiarity with series here, and you should skip Proposition 2.47 if you have not studied them in Calculus.

and each of the $(n-q)$ terms in this product is greater or equal to $q+1$ so that

$$\frac{q!}{n!} \leq \frac{1}{(q+1)^{n-q}},$$

and the inequality is strict for all $n \geq q+1$. Recall (see Theorem 2.77 for details) that the sum of a geometric series of common ratio r with $|r| < 1$ is given by

$$\sum_{i=1}^{\infty} r^i = \frac{r}{1-r}. \tag{2.8}$$

Thus,

$$x = q! \sum_{n=q+1}^{\infty} \frac{1}{n!} < \sum_{n=q+1}^{\infty} \frac{1}{(q+1)^{n-q}} \overset{i=n-q}{=} \sum_{i=1}^{\infty} \frac{1}{(q+1)^i} \underset{r=\frac{1}{q+1}<1}{\overset{(2.8)}{=}} \frac{1}{q+1} \cdot \frac{1}{1-\frac{1}{q+1}} = \frac{1}{q} < 1.$$

Thus $x \in \mathbb{Z}$ and $0 < x < 1$, which is a contradiction. Thus e is irrational. □

Proof (Proof of Theorem 2.43). We proceed by contradiction. Assume to that effect that the set of prime numbers is finite, say, there are n prime numbers p_1, $p_2 \dots p_n$ ([22]). Consider the integer

$$q = p_1 p_2 \dots p_n + 1,$$

that is, q is 1 plus the product of all n prime numbers. Because q is a natural number greater than 1, it has a prime divisor r, by Theorem 2.8. Since r is prime, $r = p_i$ for some $i \in \{1, \dots, n\}$ and thus, r divides $p_1 p_2 \dots p_n$. Thus r divides q and divides $p_1 p_2 \dots p_n$ hence, in view of Proposition 2.6, r divides $q - p_1 p_2 \dots p_n = 1$. Since r is a natural number, $r = 1$ which is a contradiction, for $r > 1$ because r is prime. ([23]) □

Additional Exercises (Homework)

Exercise 2.48. Show that the multiplicative inverse of a non-zero real number is unique.

Exercise 2.49. Show that $\sqrt{5}$ is irrational.

Exercise 2.50. Show that $\sqrt[3]{2}$ is irrational.

[22] That is $\neg P$ of (2.5).

[23] Here Q of (2.5) is either $r \neq 1$ or $r = 1$.

2.4 Special Forms of the Premises or of the Conclusion

As noted before, most statements you will have to prove are either conditional statements, or biconditional statements that you can decompose into two conditional statements, as in (1.7). Proofs of non-conditional statements may be attempted by contradiction.

In this section we examine various forms of the premise or conclusion of a conditional statement.

Premise Is a Disjunction

If the premise is a disjunction, that is, you aim to prove a statement of the form

$$(p \vee q) \Longrightarrow r,$$

you may note that

$$\begin{aligned}
(p \vee q) \Longrightarrow r &\overset{(1.4)}{\equiv} \neg(p \vee q) \vee r \\
&\overset{\text{Proposition 1.15}}{\equiv} (\neg p \wedge \neg q) \vee r \\
&\overset{(1.3)}{\equiv} (\neg p \vee r) \wedge (\neg q \vee r) \\
(p \vee q) \Longrightarrow r &\overset{(1.4)}{\equiv} (p \Longrightarrow r) \wedge (q \Longrightarrow r). \qquad (2.9)
\end{aligned}$$

In other words, to prove $(p \vee q) \Longrightarrow r$ is tantamount to proving both $p \Longrightarrow r$ and $q \Longrightarrow r$, that is, the proof of $(p \vee q) \Longrightarrow r$ can be decomposed into the proofs of two simpler conditional statements.

Example 2.51. Show that if 3 divides either n or p then 9 divides $n^2(3p - p^2)$.

Solution. In view of (2.9), it is enough to show $3|n \Longrightarrow 9|\left(n^2(3p - p^2)\right)$ and $3|p \Longrightarrow 9|\left(n^2(3p - p^2)\right)$.

$3|n \Longrightarrow 9|\left(n^2(3p - p^2)\right)$: If 3 divides n, then there is an integer k with $n = 3k$ so that

$$n^2(3p - p^2) = 9k^2(3p - p^2)$$

is a multiple of 9, that is, 9 divides $n^2(3p - p^2)$.

$3|p \Longrightarrow 9|\left(n^2(3p - p^2)\right)$: If 3 divides p, there is an integer k with $p = 3k$ so that

$$n^2(3p - p^2) = n^2(9k - 9k^2) = 9n^2(k - k^2),$$

and thus 9 divides $n^2(3p - p^2)$.

Exercise 2.52. Show that if $-2 < x < 1$ or $x > 3$ then $\dfrac{(x-1)(x+2)}{(x-3)(x+3)} > 0$.

When the **premise is a conjunction** is less interesting: both assumptions will probably be useful if you try a direct proof. On the other hand, you may instead consider the contrapositive

$$(p \wedge q) \Longrightarrow r \equiv \neg r \Longrightarrow \neg(p \wedge q) \equiv \neg r \Longrightarrow \neg p \vee \neg q, \tag{2.10}$$

which takes the form of a disjunction in the conclusion:

Conclusion Is a Disjunction

In trying to prove a statement of the form

$$p \Longrightarrow (q \vee r)$$

it is often useful to note that

$$\begin{aligned} p \Longrightarrow (q \vee r) & \overset{(1.4)}{\equiv} \neg p \vee q \vee r \\ & \overset{\text{Proposition 1.15}}{\equiv} \neg(p \wedge \neg r) \vee q \equiv \neg(p \wedge \neg q) \vee r \\ p \Longrightarrow (q \vee r) & \overset{(1.4)}{\equiv} (p \wedge \neg r) \Longrightarrow q \equiv (p \wedge \neg q) \Longrightarrow r, \end{aligned} \tag{2.11}$$

so that assuming p and trying to conclude $q \vee r$, you may assume the negation of one of q or r, and try to prove that the other is true.

Example 2.53. Show that if $x \in \mathbb{R}$ satisfies $x^2 = 2x + 15$ then $x > 2$ or $\frac{x+1}{x+2} > 0$.

Solution. Assume that $x^2 = 2x + 15$ and that $x \leq 2$. We need to show that $\frac{x+1}{x+2} > 0$.

The first assumption rephrases as

$$x^2 - 2x - 15 = 0 = (x-5)(x+3),$$

so that $x = 5$ or $x = -3$. Because we also assume $x \leq 2$, we conclude that $x = -3$, in which case

$$\frac{x+1}{x+2} = \frac{-2}{-1} = 2 > 0,$$

which concludes the proof.

Exercise 2.54. Let $x, y \in \mathbb{R}$. Prove that if $xy = 0$ then $x = 0$ or $y = 0$.

Note also that as a consequence of (2.10) and (2.11), one way to approach a proof by contraposition of a statement of the form $(p \wedge q) \Longrightarrow r$ is to note that

$$(p \wedge q) \Longrightarrow r \equiv (\neg r \wedge p) \Longrightarrow \neg q \equiv (\neg r \wedge q) \Longrightarrow \neg p, \tag{2.12}$$

that is, one may assume one of the premises and not the conclusion, and try to show the negation of the other premise.

Example 2.55. Show that if x is rational and y is irrational then $x+y$ is irrational.

Solution. The statement has the form $(p \wedge q) \Longrightarrow r$ and can thus be equivalently rephrased as $(\neg r \wedge p) \Longrightarrow \neg q$ via (2.12). In other words, it is equivalent to show that if $x+y$ is rational and x is rational, then y is rational, which is clear because if $x+y$ and x are rational, then

$$y = x + y - x$$

is rational too, because $\mathbb{Q}$ is closed under $-$.

Exercise 2.56. Show that if $x \in A$ and $x \notin A \cap B$ then $x \notin B$.

Finally, when the **conclusion is a conjunction**, say the statement has the form $p \Longrightarrow (q \wedge r)$, then you may note that

$$\begin{aligned} p \Longrightarrow (q \wedge r) &\equiv \neg p \vee (q \wedge r) \equiv (\neg p \vee q) \wedge (\neg p \wedge r) \\ p \Longrightarrow (q \wedge r) &\equiv (p \Longrightarrow q) \wedge (p \Longrightarrow r) \end{aligned} \tag{2.13}$$

and prove two conditional statements.

Additional Exercises (Homework)

Exercise 2.57. Show that if $a|b$ or $a|c$ then $a|(bc)$.

Exercise 2.58. Prove that if A, B, and C are any sets such that $A \times B = A \times C$, then either $A = \emptyset$ or $B = C$.

Exercise 2.59. Prove that for every sets A and B with $A \times B = B \times A$ then $A = B$ or $A = \emptyset$ or $B = \emptyset$.

Exercise 2.60. If n and m are positive integers for which $n \leq 2m$, then either n is prime or $m \geq \sqrt{n}$.

Exercise 2.61. Let p and b be natural numbers. Show that if p is prime and p does not divide b, then the only positive integer that divides both p and b is 1.

2.5 Disproving

Quite often when studying a problem, you formulate statements to be tested, without knowing beforehand if it is a true or a false statement. Therefore you either try to prove the statement true, or to prove it false, that is, to prove its negation. Hence to *disprove* a statement is to prove its negation. Proving or disproving a conjecture may prove extremely difficult and the reader should keep in mind that there are plenty of very important mathematical statements that remain conjectures, that is, it is unknown if they are true, false, or even independent of the usual axioms of mathematics.

2.5.1 Counterexamples

A very important activity of mathematicians is to produce what we call *counterexamples*. There are two types of statements that are disproved by a counterexample:

1. If a statement of the form
$$\forall x\, P(x) \tag{2.14}$$
is conjectured, it can be disproved by producing one example of x for which $P(x)$ is false, because in view of (1.14), $\forall x\, P(x)$ is false whenever $\exists x\, \neg P(x)$ is true. Such an x for which $P(x)$ is false is a *counterexample* to (2.14), and its existence disproves (2.14), that is, proves (2.14) false.
2. A conditional statement of the form
$$P(x) \Longrightarrow Q(x), \tag{2.15}$$
where the quantification $\forall x$ is implicit, but which really should be formulated as
$$\forall x\, (P(x) \Longrightarrow Q(x)),$$
is disproved by producing an example of x for which the conditional statement is false, that is, in view of (1.5), an example of x for which the premise $P(x)$ is true but the conclusion $Q(x)$ is false. Such an x is a counterexample to (2.15) and disproves (2.15).

Example 2.62. Prove or disprove

1. The statement $x^2 - x \geq 0$ is true for all positive real number x.
2. The statement $2^x > x+1$ is true for all positive real number x.
3. For every $n \in \mathbb{Z}$, $f(n) = n^2 - n + 11$ is prime.
4. For every triplet (A,B,C) of sets,
$$A \setminus (B \cap C) = (A \setminus B) \cap (A \setminus C).$$
5. If $x \in [0,\infty)$ and $y > 1$, then $xy > y$.
Solutions. (1) This is false, for there is $x = \frac{1}{2}$ for which
$$x^2 - x = \frac{1}{4} - \frac{1}{2} = -\frac{1}{4} < 0.$$
(2) is also false for $x = 1$ yields $2 > 2$, which is false.

(3) We check $f(n)$ for various n. It turns out that $f(0) = f(1) = 11$, $f(2) = 13$, $f(3) = 17$, $f(4) = 23$, $f(5) = 31$, $f(6) = 41$, $f(7) = 53$, $f(8) = 67$, $f(9) = 89$, $f(10) = 101$ are all prime numbers, but $f(11) = 121$ is not! Thus this is also a false statement.
(4) This is false. Take, for instance, as a counterexample $A = \{1,2,3\}$, $B = \{1\}$, and $C = \{2\}$. Then $B \cap C = \emptyset$ so that $A \setminus (B \cap C) = A = \{1,2,3\}$ while $A \setminus B = \{2,3\}$ and $A \setminus C = \{1,3\}$ so that $(A \setminus B) \cap (A \setminus C) = \{3\}$.
(5) This is false. Take $x = 0$ and $y = 2$ as a counterexample: x and y satisfy the premise that $x \in [0,\infty)$ and $y > 1$, but not the conclusion for $xy = 0 \not> 2$.

To disprove an existential statement of the form

$$\exists x \in X \;\; P(x),$$

we need to prove its negation

$$\forall x \in X \;\; \neg P(x),$$

which takes the form of a universal statement.

For instance, if you consider the conjecture

$$\exists x \in \mathbb{R} \; \left(x^4 < x < x^2\right),$$

and suspect it to be false (try to find such an x to convince yourself that it is problematic), you need to show that

$$x^4 < x < x^2$$

is false for every $x \in \mathbb{R}$. One way to do so is to proceed by contradiction.

Lemma 2.63. *For every $x \in \mathbb{R}$, it is not the case that $x^4 < x < x^2$.*

Proof. Assume to the contrary that there is $x \in \mathbb{R}$ with $x^4 < x < x^2$. Because

$$x^2 - x = x(x-1) > 0$$

then either $x < 0$ (when both factors are negative) or $x > 1$ (when both factors are positive). Moreover,

$$x^4 - x = x(x^3 - 1) < 0.$$

If $x < 0$ and $x(x^3 - 1) < 0$, then $x^3 - 1 > 0$ so that $x > 1$. Hence $x < 0$ and $x > 1$ which is a contradiction.

On the other hand, if $x > 1$ and $x(x^3 - 1) < 0$, then $x < 0$ and we obtain the same contradiction. □

Exercise 2.64. Prove or disprove

1.

$$\forall x, y \in \mathbb{R}, \; |x+y| = |x| + |y|.$$

2. If $n \in \mathbb{Z}$ and $n^5 - n$ is even, then n is even.

3. If A, B, C are sets such that $A \times C = B \times C$, then $A = B$.
4. Let $|X|$ denote the number of elements of a finite set X. If X and Y are finite, then $|X \cup Y| = |X| + |Y|$.

Additional Exercises (Homework)

Exercise 2.65. Prove or disprove

1. There is a natural number N such that $\frac{1}{n} < N$ for all $n \in \mathbb{N}$.
2. For every triplet (A, B, C) of sets

$$A \setminus (B \cup C) = (A \setminus B) \cup (A \setminus C).$$

3. If $n, p \in \mathbb{N}$, then $n + p < np$.
4. For every pair (X, Y) of sets, if $X \setminus Y = \emptyset$ then $Y \neq \emptyset$.
5. For every pair (X, Y) of sets,

$$\mathbb{P}X \cap \mathbb{P}Y = \mathbb{P}(X \cap Y).$$

6. If $C \subset A \cup B$, then $C \subset A$ or $C \subset B$.
7. For every real numbers x and y

$$|x + y| = |x - y| \Longrightarrow y = 0.$$

8. For all positive integer n, the number $n^2 + n + 41$ is prime.
9. For all integers a, b, and c

$$a|(bc) \Longrightarrow (a|b) \vee (a|c).$$

2.6 Proof by Induction

How do we prove that something that depends on a natural number n is true *for all* $n \in \mathbb{N}$?

For instance, you may want to know what is the sum of the first n odd natural numbers

$$\sum_{i=1}^{n} (2i - 1) = 1 + 3 + 5 \ldots + (2n - 1),$$

as a function of n. To find out, you would start by looking at what happens for the first few values of n:

$$n=1 \Longrightarrow \sum_{i=1}^{n}(2i-1)=1$$

$$n=2 \Longrightarrow \sum_{i=1}^{n}(2i-1)=1+3=4$$

$$n=3 \Longrightarrow \sum_{i=1}^{n}(2i-1)=1+3+5=9$$

$$n=4 \Longrightarrow \sum_{i=1}^{n}(2i-1)=1+3+5+7=16$$

$$\vdots$$

You may note that for $n=1$, $\sum_{i=1}^{n}(2i-1)=1^2$, for $n=2$, $\sum_{i=1}^{n}(2i-1)=2^2$, for $n=3$, $\sum_{i=1}^{n}(2i-1)=3^2$, and you may try a few more values of n and see that the formula

$$\sum_{i=1}^{n}(2i-1)=n^2 \tag{2.16}$$

seems to be true *for all* $n \in \mathbb{N}$. The problem is that you only verified it for a few values of n, and (2.16) is really an infinite set of formulas: one for each $n \in \mathbb{N}$. That you verified that the formula is true for the first few n's is definitely no proof that it remains true for all n!

But how to verify infinitely many instances?

The "trick" is to use *mathematical induction* or *induction* for short. The idea is the following: To prove

$$\forall n \in \mathbb{N}\, P(n),$$

it is enough to show that

1. $P(1)$ is true (*base case*: verify $P(n)$ for $n=1$);
2. Show that for every $n \in \mathbb{N}$,

$$P(n) \Longrightarrow P(n+1). \tag{2.17}$$

The second step is often called the *inductive step*, in which we make the *inductive hypothesis* that $P(n)$ is true. In this step we prove that **if** $P(n)$ is true for a certain n **then** $P(n+1)$ is necessarily also true.

How does that suffices to conclude that $P(n)$ is true for **all** n? Well, we checked $P(1)$. But according to the inductive step for $n=1$, if $P(1)$ is true, then $P(2)$ is true. Thus the inductive step can be applied with $n=2$: if $P(2)$ is true, then $P(3)$ is true too, which means that the inductive step applies again to the effect that $P(4)$ is true, and so on. The inductive step (2.17) is what guarantees a domino effect to get all instances of $P(n)$ true, as long as it is true in the base case, for $n=1$.

We clarify what property of the natural numbers makes this work in the Appendix A.3.

Let us use this technique to prove (2.16). In this case $P(n)$ is

$$\sum_{i=1}^{n}(2i-1)=n^2. \tag{2.18}$$

Proof (Proof of (2.16)). If $n=1$, $\Sigma_{i=1}^{1}(2i-1)=1=1^2$ so $P(1)$ is true.

Assume that $n\geq 1$ is an integer such that $P(n)$, that is, such that $\Sigma_{i=1}^{n}(2i-1)=n^2$. We need to prove that $P(n+1)$ is true, that is,

$$\sum_{i=1}^{n+1}(2i-1)=(n+1)^2, \tag{2.19}$$

which is obtained by replacing each instance of n by $(n+1)$ in $P(n)$.

To this end, note that, taking the last term of the sum aside (and taking into account Remark 2.66 below), we have:

$$\begin{aligned}\sum_{i=1}^{n+1}(2i-1) &= 1+3+\ldots(2n-1)+2(n+1)-1\\ &= \sum_{i=1}^{n}(2i-1)+2n+1\\ &= n^2+2n+1\end{aligned}$$

because by inductive assumption $\Sigma_{i=1}^{n}(2i-1)=n^2$. Thus

$$\sum_{i=1}^{n+1}(2i-1)=n^2+2n+1=(n+1)^2.$$

It follows by induction that (2.16) is true for all $n\in\mathbb{N}$. □

Remark 2.66. Note that in proving (2.19) under the assumption that (2.16) is true for n, we do not work from both sides of (2.19). Rather, we start from one side and use the inductive assumption to obtain the other side. Do **not** work on the complete formula (2.19) as it is, for in doing so, you would be taking as a starting point, hence an assumption, the very thing you want to establish!

Proposition 2.67.

$$\sum_{i=1}^{n} i=\frac{1}{2}n(n+1), \tag{2.20}$$

for every $n\in\mathbb{N}$.

We consider two different proofs of this fact. A "direct" proof that also gives you the formula if you didn't know it, and is simple once you know the "trick"—yet it requires a trick—and a proof by induction that requires that you know what to prove, but is a routine exercise if you know the formula.

Proof (Direct Proof of Proposition 2.67). Let n be a natural number and let $s_n = \sum_{i=1}^{n} i$. We write s_n in two ways, by writing the terms in ascending, then in descending fashion, and add "column by column"

$$\begin{array}{rcccc} s_n = & 1+2+ & 3+\ldots+ & n-1+n \\ s_n = & n+n-1+ & n-2+\ldots+ & 2+1 \\ 2s_n = & (n+1)+(n+1)+ & (n+1)+\ldots+ & (n+1)+(n+1), \end{array}$$

so that $2s_n = n(n+1)$, because each of the n columns adds up to $n+1$. Therefore

$$\sum_{i=1}^{n} i = \frac{1}{2}n(n+1).$$

□

Remark 2.68. You can interpret this approach geometrically:

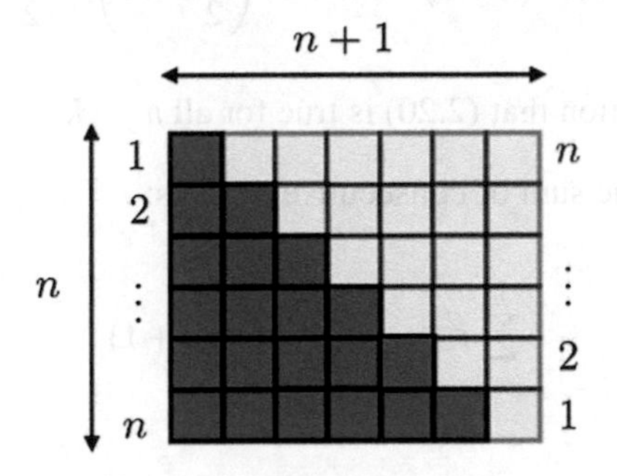

In this picture, you can think of the dark squares as the first version of s_n where the terms are in ascending order top to bottom (1 square, then 2, then 3, etc.) and the light squares represent the second version where the terms are in descending order top to bottom (n squares, then $n-1$, etc.). Together, they form an $n \times (n+1)$ rectangle that represents twice the sum we are looking for.

Proof (Proof of Proposition 2.67 by Induction). We prove that (2.20) for all $n \in \mathbb{N}$ by induction. To this end, we first verify that this is true for $n = 1$: $\sum_{i=1}^{1} i = 1 = \frac{1}{2} \times 1 \times 2$. To prove the inductive step, assume that

$$\sum_{i=1}^{n} i = \frac{1}{2}n(n+1)$$

for some $n \in \mathbb{N}$. We need to show that

$$\sum_{i=1}^{n+1} i = \frac{1}{2}(n+1)(n+2),$$

which is obtained from the formula to be proven by replacing every instance of n by $n+1$.

To this end (keeping Remark 2.66 in mind) note that

$$\sum_{i=1}^{n+1} i = 1+2++\ldots+n+(n+1) = \sum_{i=1}^{n} i + (n+1),$$

by taking the last term of the sum aside. Moreover, $\sum_{i=1}^{n} i = \frac{1}{2}n(n+1)$ by inductive hypothesis. Hence

$$\sum_{i=1}^{n+1} i = \frac{1}{2}n(n+1) + (n+1) = (n+1)\left(\frac{1}{2}n+1\right) = \frac{1}{2}(n+1)(n+2).$$

We conclude by induction that (2.20) is true for all $n \in \mathbb{N}$. □

Let us now consider the sum of consecutive squares:

Proposition 2.69.

$$\sum_{i=1}^{n} i^2 = \frac{1}{6}n(n+1)(2n+1) \tag{2.21}$$

for every $n \in \mathbb{N}$.

Proof. We prove (2.21) for all $n \in \mathbb{N}$ by induction. We verify first that (2.21) is true for $n = 1$:

$$\sum_{i=1}^{1} i^2 = 1 = \frac{1}{6} \times 1 \times 2 \times 3.$$

Assume now that (2.21) is true for some $n \in \mathbb{N}$. We need to show that (2.21) is true for $n+1$, that is, that

$$\begin{aligned}\sum_{k=1}^{n+1} k^2 &= \frac{1}{6}(n+1)(n+1+1)(2(n+1)+1) \\ &= \frac{1}{6}(n+1)(n+2)(2n+3).\end{aligned}$$

To see this, note that (proceeding along the same lines as for the previous two examples)

$$\begin{aligned}\sum_{k=1}^{n+1} k^2 &= \sum_{k=1}^{n} k^2 + (n+1)^2 \overset{\text{ind. hyp.}}{=} \frac{n(n+1)(2n+1)}{6} + (n+1)^2 \\ &= \frac{1}{6}\left(n(n+1)(2n+1) + 6(n+1)^2\right) = \frac{1}{6}\left((n+1)\left(2n^2+n+6n+6\right)\right) \\ &= \frac{1}{6}\left((n+1)(2n^2+7n+6)\right) = \frac{1}{6}(n+1)(n+2)(2n+3).\end{aligned}$$

We conclude by induction that (2.21) is true for all $n \in \mathbb{N}$. □

Exercise 2.70. Conjecture a formula for the sum of the first n consecutive even integers, and prove it by induction.

Remark 2.71. Of course, if you want to prove that a statement $P(n)$ that depends on an integer n is true for every $n \geq k$ for some fixed $k \in \mathbb{Z}$, we can use induction in the exact same way except that the base case would be to verify the case $n = k$ instead of $n = 1$. The "domino effect" induced by

$$P(n) \Longrightarrow P(n+1) \tag{2.22}$$

remains unchanged, and if we have the result for $n = k$, then we have it for $n = k+1$ by (2.22), and thus for $k+2$ using (2.22) again, and so on and so forth.

Proposition 2.72. *Show that if n is a non-negative integer, then $5|(n^5 - n)$.*

Note that this is not a transparent statement when we do not know what the value of n is. But as long as we know what to prove, induction is a very powerful tool:

Proof. We proceed by induction. If $n = 0$, then $n^5 - n = 0$ and any number divides 0; in particular, $0 = 0 \times 5$, so $5|0$.

Assume now that $5|(n^5 - n)$ for a certain non-negative integer n. We need to show that

$$5|((n+1)^5 - (n+1)).$$

By the hypothesis of induction, there is an integer k with

$$n^5 - n = 5k. \tag{2.23}$$

In view of Theorem 1.120,

$$\begin{aligned}(n+1)^5 - (n+1) &= n^5 + 5n^4 + 10n^3 + 10n^2 + 5n + 1 - (n+1) \\ &= (n^5 - n) + 5(n^4 + 2n^3 + 2n^2 + n) \\ &\overset{(2.23)}{=} 5k + 5(n^4 + 2n^3 + 2n^2 + n) = 5(n^4 + 2n^3 + 2n^2 + n + k)\end{aligned}$$

is thus a multiple of 5, that is, $5|((n+1)^5 - (n+1))$. We conclude by induction that $5|(n^5 - n)$ for every integer $n \geq 0$. □

Exercise 2.73. Show that $3|(n^3-n)$ for all non-negative integer n.

Example 2.74. Show that $2^n < n!$ for all integer $n \geq 4$.

Solution. We proceed by induction. For $n=4$, $2^n = 2^4 = 16$ and $n! = 4! = 24$ so $2^n < n!$.

Assume now that $2^n < n!$ for a certain $n \geq 4$. We need to show that $2^{n+1} < (n+1)!$. To this end, note that

$$2^{n+1} = 2^n \cdot 2 < 2 \cdot n! \tag{2.24}$$

by the hypothesis of induction. It remains to be seen that $2 \cdot n! \leq (n+1)!$. To this end, note that $n \geq 4$ so that $2 < 4 < n+1$ and thus

$$2 \cdot n! < (n+1) \cdot n! = (n+1)!,$$

which, combined with (2.24), yields $2^{n+1} < (n+1)!$ as desired. We conclude by induction that $2^n < n!$ for all integer $n \geq 4$.

Exercise 2.75. Show that

$$n+3 < 5n^2$$

for all $n \in \mathbb{N}$.

Exercise 2.76. Show that

$$2^n \leq 2^{n+1} - 2^{n-1} - 1$$

for every $n \in \mathbb{N}$.

Theorem 2.77. *Let $n \in \mathbb{N}$. The sum of n consecutive terms of a geometric series of common ratio $r \neq 1$ and first term a is given by*

$$\sum_{i=0}^{n-1} ar^i = a\frac{(r^n-1)}{r-1}. \tag{2.25}$$

Proof (Proof by Induction). For $n=1$, $\sum_{i=0}^{n-1} ar^i = ar^0 = a$ and $a\frac{(r^n-1)}{r-1} = a\frac{r-1}{r-1} = a$, so the formula is true.

Assume now that (2.25) for some integer $n \geq 1$. We need to show that

$$\sum_{i=0}^{n} ar^i = a\frac{(r^{n+1}-1)}{r-1}.$$

To this end, note that

$$\sum_{i=0}^{n} ar^i = a + ar + ar^2 + \ldots ar^{n-1} + ar^n = \sum_{i=0}^{n-1} ar^i + ar^n$$

and by hypothesis of induction $\sum_{i=0}^{n-1} ar^i = a\frac{(r^n-1)}{r-1}$. Thus

$$\sum_{i=0}^{n} ar^i = a\frac{(r^n-1)}{r-1} + ar^n = a\frac{(r^n-1)+r^n(r-1)}{r-1}$$
$$= a\frac{r^n-1+r^{n+1}-r^n}{r-1} = a\frac{r^{n+1}-1}{r-1}.$$

We conclude by induction that (2.25) is true for all $n \in \mathbb{N}$. □

Note that proving this formula by induction presupposes that we know the formula to start with. If you do not know the formula, a trick similar to that used for the direct proof of Proposition 2.67 can be used:

Proof (Alternative "Direct" Proof of Theorem 2.77). Let $s_n = \sum_{i=0}^{n-1} ar^i$. Note that writing s_n and rs_n and subtracting "in columns," we have

$$\begin{array}{llll} s_n = a+ & ar+ar^2+ & \ldots+ar^{n-2}+ & ar^{n-1} \\ rs_n = & ar+ar^2+ & \ldots+ar^{n-2}+ & ar^{n-1}+ar^n \\ s_n - rs_n = a & & & -ar^n, \end{array}$$

so that

$$s_n(1-r) = a(1-r^n)$$

and, since $r \neq 1$, we get by division

$$s_n = \frac{a(1-r^n)}{1-r},$$

that is, we obtain (2.25). □

Exercise 2.78. Show that

$$\sum_{i=1}^{n} 2^i = 2^{n+1} - 2.$$

Exercise 2.79. Show that

$$\sum_{i=0}^{n} i \cdot i! = (n+1)! - 1$$

for every integer $n \geq 0$.

Example 2.80. Show that

$$\prod_{i=1}^{n}(2i-1) = \frac{(2n)!}{n!2^n}, \tag{2.26}$$

for all natural number n.

Solution. For $n = 1$, the product $\prod_{i=1}^{n}(2i-1) = 2-1 = 1$ and $\frac{(2n)!}{n!2^n} = \frac{2!}{1!2} = \frac{2}{2} = 1$, hence the formula is verified.

Assume now that (2.26) is verified for some natural number n. We want to show that it also holds for $n+1$, that is, we want to show that

$$\prod_{i=1}^{n+1}(2i-1) = \frac{(2(n+1))!}{(n+1)!2^{n+1}} = \frac{(2n+2)!}{(n+1)!2^{n+1}}. \tag{2.27}$$

To this end, we note that taking the factor corresponding to $i = n+1$ aside,

$$\begin{aligned}\prod_{i=1}^{n+1}(2i-1) &= (2(n+1)-1)\prod_{i=1}^{n}(2i-1)\\ &= (2n+1)\prod_{i=1}^{n}(2i-1),\end{aligned}$$

and we use the inductive hypothesis to the effect that

$$\prod_{i=1}^{n+1}(2i-1) = (2n+1)\frac{(2n)!}{n!2^n}.$$

Rewriting

$$2n+1 = \frac{(2n+1)(2n+2)}{2(n+1)},$$

we obtain

$$\begin{aligned}\prod_{i=1}^{n+1}(2i-1) &= \frac{(2n+1)(2n+2)}{2(n+1)}\cdot\frac{(2n)!}{n!2^n}\\ &= \frac{(2n)!(2n+1)(2n+2)}{n!(n+1)\cdot 2^n\cdot 2}\\ &= \frac{(2n+2)!}{(n+1)!2^{n+1}},\end{aligned}$$

as in (2.27). We conclude by induction that (2.26) is true for all $n \in \mathbb{N}$.

The following exercise illustrates the care needed in considering the base case, and how the inductive step articulates with the base case:

Exercise 2.81. Here is a "proof" that any set of billiard balls is made of balls of the same color. As this is a transparently false statement, there must be an error in the proof. Find it!

We prove by induction the statement $P(n)$ given by "any group of n billiard balls is formed by balls of the same color," for every $n \in \mathbb{N}$. For $n = 1$, the statement is true. Assume that this is true for a certain n and consider a group of $n+1$ balls. Take one ball aside. The n remaining balls are all of the same color by the inductive assumption. Put back the ball placed aside and remove another ball, resulting in a group of n balls. By the inductive assumption, they are all of the same color, so that the ball set aside first was of the same color as the others. Hence $P(n+1)$ is true.

Let us examine an alternative proof of Corollary 1.116 by induction:

Proof (Proof of Corollary 1.116 by Induction). We want to show by induction on an integer $n \geq 0$ that if X has n elements then $\mathbb{P}X$ has 2^n elements. If $n = 0$, then $X = \emptyset$ and $\mathbb{P}X = \{\emptyset\}$ has $2^0 = 1$ element. Hence this is true for $n = 0$.

Assume now that for some integer $n \geq 0$, every set with n elements has a powerset with 2^n elements. We need to show that if a set has $n+1$ elements, then its powerset has 2^{n+1} elements. To this end, let X have $n+1$ elements and let $x_0 \in X$ ([24]) be a particular element. Then $X \setminus \{x_0\}$ has n elements, hence has 2^n subsets by inductive hypothesis ([25]). But for each $A \subset X$, either $x_0 \notin A$, that is, $A \in \mathbb{P}(X \setminus \{x_0\})$, or $x_0 \in A$ and $A = (A \setminus \{x_0\}) \cup \{x_0\}$. Hence there are 2^n subsets of X that do not contain x_0 and 2^n that do. In other words, there are

$$2^n + 2^n = 2^{n+1}$$

subsets of X. □

Proposition 2.82 (Bernoulli's Inequality). *If $n \in \mathbb{N}$ then*

$$(1+x)^n \geq 1 + nx$$

for every real number $x > -1$.

Note that here the statement we want to prove valid for every $n \in \mathbb{N}$ depends both on n and x, that is, has the form $P(n,x)$. This is no problem as long as you are clear on what variable the induction is carried on.

Proof. For $n = 1$, $(1+x)^1 = 1+x$ and $1+nx = 1+x$, so we have equality for all x, in particular for all $x > -1$.

Assume now that for a certain natural number n, $(1+x)^n \geq 1+nx$ for all $x > -1$. We need to show that

$$(1+x)^{n+1} \geq 1 + (n+1)x$$

for all $x > -1$. To this end, note that

$$(1+x)^{n+1} = (1+x)^n \cdot (1+x),$$

so that, using the inductive hypothesis ([26]),

$$\begin{aligned}(1+x)^{n+1} &\geq (1+nx)\cdot(1+x)\\ &\geq 1+(n+1)x+nx^2\\ &\geq 1+(n+1)x\end{aligned}$$

for $nx^2 \geq 0$. We conclude by induction that $(1+x)^n \geq 1+nx$ for every $x > -1$. □

[24] Note that X has $n+1$ elements for $n \geq 0$, so $X \neq \emptyset$.

[25] Note that here n might be 0, hence 2^n could be $2^0 = 1$.

[26] Note that the direction of the inequality is preserved by multiplication by $(1+x)$ because we have assumed $x > -1$, that is, $x+1 > 0$.

2.6.1 Smallest Counterexample

Sometimes, to prove a certain statement $P(n)$, we verify the base case and we prove the inductive step *by contradiction*, in the following sense: we assume that it is not true that $P(n)$ is true for all n, that is, there is k with $P(k)$ false. We can then assume that k is the *smallest such* k (that is, the first k for which $P(k)$ fails). In particular, $P(k-1)$ is then true. We seek a contradiction from

$$P(k-1) \wedge \neg P(k).$$

If we can obtain a contradiction, then the assumption that it is not true that $P(n)$ for all n is false, that is, $P(n)$ is true for all n. This technique is sometimes referred to as *a proof by smallest counterexample* (because we use the fact that if there is a counterexample n to "$\forall n\, P(n)$," there is also the smallest one).

To illustrate this technique, consider:

Proposition 2.83. *For every* $n \in \mathbb{N}$, $4|(5^n - 1)$.

Proof (Proof by Smallest Counterexample). This is true for $n = 1$ for 4 divides $5^1 - 1 = 4$.

Assume by way of contradiction that there is some integer $n > 1$ for which $4 \not|(5^n - 1)$. We may assume that n is the smallest such n. In particular, $4|(5^{n-1} - 1)$, that is, there is an integer k for which

$$5^{n-1} - 1 = 4k,$$

so that

$$5(5^{n-1} - 1) = 5 \times 4k = 20k,$$

that is,

$$5^n - 5 = 20k \Longrightarrow 5^n - 1 = 20k + 4 = 4(5k+1)$$

and thus $4|(5^n - 1)$, contradicting the assumption. □

Note that we could easily proceed more straightforwardly, as long as you know the factorization formula

$$a^n - b^n = (a-b)(a^{n-1} + a^{n-2}b + a^{n-3}b^2 + \ldots + ab^{n-2} + b^{n-1}) \qquad (2.28)$$

for every real numbers a, b and natural number n.

Exercise 2.84. Prove (2.28).

Proof (Proof Using Factorization). Let $n \in \mathbb{N}$. In view of (2.28) for $a = 5$ and $b = 1$,

$$5^n - 1 = 5^n - 1^n = (5-1)(5^{n-1} + 5^{n-2} + \ldots + 1),$$

so that $4|(5^n - 1)$. □

2.6.2 A Detour on the Well-Ordering Principle and Arithmetic

Note that the argument above that if there is a k for which $P(k)$ is false there is also the smallest such k relies on the fact that *every non-empty subsets of* $\mathbb{N}$ *has a smallest element*. This is known as the *well-ordering principle*. In this subsection, we consider some important examples of proofs in arithmetic that rely on the well-ordering principle. Formally, they do not involve induction, and thus one may argue that they do not belong here. However, the well-ordering principle and the principle of induction are two sides of the same coin, as shown in Theorem A.9 of Appendix A. In other words, a proof relying on the well-ordering principle is essentially a proof by induction in disguise.

Theorem 2.85 (Euclidean Division). *Given integers a and $b \neq 0$, there exist unique integers q and r such that*

$$a = bq + r \text{ and } 0 \leq r < |b|.$$

The number q is called the quotient *and the number r is called the* remainder *of the division of a by b.*

Proof. We first show existence, starting with the case where $a \geq 0$ and $b > 0$. If $b > a$ then $a = 0 \cdot b + a$ and $r = a$ satisfies $0 \leq r < b$. Assume now that $b \leq a$. The set $S = \{a - bn : n \in \mathbb{N} \wedge a - bn \geq 0\}$ is a non-empty subset of $\mathbb{N} \cup \{0\}$ (because $a - b \in S$) and thus has a smallest element r, by the well-ordering principle. There is $q \in \mathbb{N}$ for which $r = a - bq$, that is, $a = bq + r$. Remains to see that $r < b$. If, on the contrary, $r = a - bq \geq b$, then $r' = a - b(q+1) \geq 0$ so that $r' \in S$, and $r' < r$. Therefore r would not be the smallest element of S.

To complete the proof of existence of a pair (q, r) of integers as desired for every integers a and $b \neq 0$, it remains to consider the cases $b < 0$ and $a < 0$, $b < 0$ and $a \geq 0$, and the case $a < 0$ and $b > 0$, each of which reduces to the first case (see Exercise 2.86 below).

To show uniqueness, suppose that there exists q, q' and r, r' with

$$a = bq + r = bq' + r' \tag{2.29}$$

with $0 \leq r, r' < |b|$. Then $-|b| < r - r' < |b|$ so that $|r - r'| < |b|$ ([27]). On the other hand, subtracting the two forms of a in (2.29) yields $b(q - q') = r' - r$, so that $|b|$ divides $|r - r'|$. In view of $|r - r'| < |b|$, we conclude that $|r - r'| = 0$, that is, $r = r'$. Hence $b(q - q') = 0$ and $b \neq 0$, so that $q = q'$. □

---

[27] Indeed, as by definition

$$|a| = \begin{cases} a & \text{if } a \geq 0 \\ -a & \text{if } a < 0 \end{cases},$$

we have

$$-|a| < x < |a| \iff |x| < |a|.$$

Exercise 2.86. Complete the proof of Theorem 2.85 by showing that the cases $b<0$ and $a<0$, $b<0$ and $a\geq 0$, and $a<0$ and $b>0$ can be treated from the case $a\geq 0$ and $b>0$.

An important corollary of the Euclidean division theorem whose proof also uses the well-ordering principle is:

Theorem 2.87 (Bezout's Identity). *Let a and b be non-zero integers and let $d=\gcd(a,b)$ be their* greatest common divisor ([28]). *Then there are integers x and y such that*

$$ax+by=d.$$

Moreover, d is the smallest positive integer of the form $ax+by$.

Proof. Let $S=\{ax+by : x,y\in\mathbb{Z}\}$. Obviously, $S\cap\mathbb{N}\neq\emptyset$ so that $S\cap\mathbb{N}$ has a smallest element s (by the well-ordering principle). Every element n of S is divisible by s. Assume to the contrary that there is $n\in S$ where s does not divide n. Then the Euclidean division of n by s yields integers q and r with

$$n=qs+r,\ 0<r<s.$$

Then $r=n-qs$ and n and s both belong to S, so that there are integers x,y with $n=ax+by$ and integers k and ℓ with $s=ka+\ell b$. Thus

$$r=ax+by-q(ka+\ell b)=(x-qk)a+(b-q\ell)b\in S,$$

in contradiction with $0<r<s$ and the definition of s as the smallest positive element of S.

Since a and b belong to S (take $(x,y)=(1,0)$ and $(x,y)=(0,1)$, respectively), we conclude that s is a common divisor of a and b. On the other hand, if c is another common divisor of a and b, then $a=pc$ and $b=mc$ so that

$$s=ka+\ell b=(kp+\ell m)c.$$

Thus $c|s$ so that $s\geq c$ and $s=\gcd(a,b)$. □

Corollary 2.88 (Euclid's Lemma). *If a prime number p divides a product ab, then p divides a or p divides b.*

Note that an immediate induction yields that if a prime number p divides a product (of finitely many factors), then p divides one of the factors.

Proof. In view of (2.11), we may assume that p is prime and that p does not divide a, and show that p divides b. If p does not divide a, the greatest common divisor of p and a is 1, so that in view of Theorem 2.87, there are x and y such that

$$px+ay=1,$$

---

[28] That is, d is a common divisor of a and b and additionally,

$$(n|a\wedge n|b)\Longrightarrow n\leq d.$$

so that, multiplying by b,

$$pxb + yab = b.$$

Since p divides pxb and divides yab because p divides ab, we conclude that p divides their sum b. □

Example 2.89. If a is an integer such that 7 divides $3a$, then by Corollary 2.88, $7|3$ or $7|a$ because 7 is prime. As 7 does not divide 3, we conclude that 7 divides a. In other words,

$$7|3a \Longrightarrow 7|a.$$

2.6.3 Strong Induction

Sometimes, to show the inductive step, it is not enough to assume $P(n)$ to obtain $P(n+1)$, but

$$(P(1) \wedge P(2) \wedge \ldots \wedge P(n)) \Longrightarrow P(n+1) \qquad (2.30)$$

might do the trick. In other words, you may need to assume that $P(k)$ is true for all k up to a certain point n to show that the next n works, that is, to show $P(n+1)$. In such a case, you typically (but not always) need to verify $P(1)$, $P(2)$, and possibly several other instances of $P(i)$ in the initial step, depending on how many of the $P(i)$ for $i \leq n$ you use, or how far before $P(n)$ you need to reach, in showing (2.30). This is called *strong induction*.

Example 2.90. Consider the following

Claim 2.91. Every amount of postage that is at least 12 cents can be made from 4-cent and 5-cent stamps.

For instance, 12 cents can be made with three stamps of 4-cent. 13 cents of postage can be obtained from two 4-cent stamps and a 5-cent stamp. The basic idea of the induction is that if you know you can do it for $n-4$ cents of postage then you can add a 4-cent stamp to make it work for n cents. But that means you need to "reach back" 4 steps. In such a case, you'd have to check the first 4 instances of your property. Here we want to show

$P(n)$: "a postage of n cents can be made from 4-cent and 5-cent stamps"

for every $n \geq 12$. So our base case is going to be to verify $P(12)$, $P(13)$, $P(14)$ and $P(15)$. Let us write this down more formally:

Proof. We prove Claim 2.91 by (strong) induction. As a base case, we verify that postage of 12, 13, 14, or 15 cents can be obtained with 4-cent and 5-cent stamps. Indeed, 12 cents are obtained with three 4-cent stamps, 13 cents are obtained with two 4-cent stamps and a 5-cent stamp, 14 cents are obtained with two 5-cent stamps and a 4-cent stamp, and 15 cents are obtained with three 5-cent stamps ([29]).

Assume now that for every integer k between 12 and n, we can obtain k cents of postage with 4-cent and 5-cent stamps. We need to show that this can be done with $n+1$ cents of postage, where $n+1$ can be assumed to be at least 16, for we have already proved the result up to 15 cents. Since $n+1 \geq 16$, $k = (n+1)-4 \geq 12$ and thus $12 \leq k \leq n$. By the hypothesis of induction, k cents of postage can be obtained with 4-cent and 5-cent stamps. Adding a 4-cent stamp then yields $k+4 = n+1$ cents of postage. □

Note that if you are going to use strong induction, it is often hard to tell beforehand how many base cases you are going to need until you work out the details of the inductive steps to see how far back you are going to reach.

Example 2.92 (Game of Nim). Nim is the name of a game with two or more players taking turns removing objects (say, matches) from (any number of) distinct piles. On each turn, a player must remove at least one object and may remove any number of objects as long as they all come from the same heap (but it can be from any heap). The goal is to be the player to remove the last object ([30]).

Claim. If two players play Nim with two piles that contain the same number of objects at the start of the game, then the second player has the following winning strategy: If the first player removes m objects from one pile, the second player removes m objects from the other heap.

Proof. We prove by (strong) induction on the number n of objects in each heap that this is indeed a winning strategy for player II.

We verify the case $n = 1$: the first player can only remove the only object in one of the pile, and the second player wins by taking the one object in the second pile.

Assume now by induction that player II wins for every game starting with two heaps of k objects, for all $1 \leq k \leq n$. We need to show that player II wins if we start with two heaps of $n+1$ objects. A move by player I is to remove j objects from one of the pile, where $1 \leq j \leq n+1$. After this move, one pile contains $n+1$ objects, and the other $k = n+1-j$. Note that $0 \leq k \leq n$. If player II removes j objects from the other pile, the second pile has also k objects. If $k = 0$, player II has won (he already removed the last object). If $1 \leq k \leq n$, then by the inductive hypothesis, player II wins because it is player I's turn and we start with two piles each of which contains k objects for $1 \leq k \leq n$. □

Note that here we do not need a base case that involves more than one verification. Indeed, we reach back an arbitrary number of steps. Hence when $n = 2$, we can reach back to the case $n = 0$ (trivial) or $n = 1$ (base case). When $n = 3$, we can reach back to $n = 2$ (already obtained from $n = 1$), $n = 1$ (base), or $n = 0$ (trivial), and similarly for larger n.

---

[29] This completes the base case of the induction.

[30] See, e.g., the wikipedia entry on "Nim" for more details.

Theorem 2.93 (Fundamental Theorem of Arithmetic). *Every integer $n \geq 2$ has a (unique up to order) prime factorization, that is, n can be expressed as a product of one or more prime numbers (in a unique fashion up to the order of factors).*

We show by strong induction on n that n has a prime factorization.

Proof. When $n = 2$, the integer n has a (unique) prime factorization with unique prime factor 2. Assume now that every k with $2 \leq k \leq n$ has a prime factorization. We need to show that $n+1$ has a prime factorization. If $n+1$ is prime, we are done. Else, $n+1 = ab$ with $1 < a < n+1$ and $1 < b < n+1$. The inductive hypothesis applies to both a and b to the effect that they are both product of prime numbers. Hence, so is $ab = n+1$.

We now prove that the factorization is unique (up to the order of factors) by using an argument by smallest counterexample. By way of contradiction, assume that n has two different factorizations

$$n = p_1 \cdot p_2 \cdot \ldots p_k = a_1 \cdot a_2 \ldots a_\ell.$$

We may assume as well that n is the smallest integer with two factorizations, for if there is such an integer, there exists also the smallest one. Since p_1 is a prime factor of $n = a_1 \cdot \ldots a_\ell$, then in view of Euclid's Lemma (Corollary 2.88) p_1 divides one of the a_i's for some $i \in \{1, \ldots \ell\}$. Since a_i is prime, $a_i = p_1$. Dividing

$$n = p_1 \cdot p_2 \cdot \ldots p_k = a_1 \cdot a_2 \ldots a_\ell$$

by $p_1 = a_i$ yields

$$p_2 \cdot \ldots p_k = a_1 \cdot a_2 \ldots a_{i-1} \cdot a_{i+1} \ldots a_\ell. \tag{2.31}$$

These two prime factorizations are different for the two factorizations of n were assumed different and we only simplified a common factor. But the number (2.31) is smaller than n and has two different factorizations, in contradiction to the definition of n. □

Remark 2.94. Note that Theorems 2.85 and 2.93 both state the existence and uniqueness of certain integers satisfying certain identities. In both cases, the uniqueness part is showed along the line of the uniqueness part in (1.11): if there are two, then they are the same:

$$P(y) \wedge P(z) \Longrightarrow y = z.$$

In the proof of Theorem 2.85, we prove this conditional statement in a direct fashion. In the proof of Theorem 2.93, we prove this conditional statement by showing that $P(y) \wedge P(z) \wedge (y \neq z)$ leads to a contradiction. You should remember that the typical proof of uniqueness starts with the assumption that you have two solutions. You then proceed to show either that they are actually the same or that if they are different we can obtain a contradiction.

2.6.4 Fibonacci Numbers

Let $\{F_n\}_{n=1}^{\infty}$ be the sequence defined inductively by

$$\begin{cases} F_1 = F_2 = 1 \\ F_{n+1} = F_n + F_{n-1} \quad \text{for every } n \geq 2. \end{cases} \tag{2.32}$$

In other words,

$$F_3 = F_2 + F_1 = 1 + 1 = 2,$$
$$F_4 = F_3 + F_2 = 2 + 1 = 3,$$

and so on. Namely,

$$F_1 = 1, F_2 = 1, F_3 = 2, F_4 = 3, F_5 = 5, F_6 = 8, F_7 = 13, F_8 = 21, F_9 = 34, \ldots$$

This is the sequence of *Fibonacci numbers*, named after the 12th century Italian mathematician Leonardo Pisano, known as Fibonacci, whose book *Liber Abaci* is credited with popularizing the Hindu-Arabic number system in medieval Europe.

Many sequences can be defined *inductively*, that is, by giving the first term (or first few terms) and an inductive rule giving the next term in the sequence in function of the previous (or few previous) terms. For instance, the sequence $x_n = n!$ defining factorials can be defined inductively as

$$\begin{cases} x_1 = 1 \\ x_{n+1} = (n+1) \cdot x_n \quad \text{for all } n \in \mathbb{N} \end{cases}.$$

Properties of sequences defined inductively are often "easy" to prove by induction, precisely because they are defined through an inductive process, that is, because one term only depends on the previous term or terms. In this section, we illustrate this fact with various properties of Fibonacci numbers. These numbers satisfy many interesting identities, and we will just show a few. We should emphasize that once the identity is found, it usually can be proved by induction. The difficult part is in conjecturing the right formula by identifying a pattern in the first few terms.

Proposition 2.95. *The sum of the first n Fibonacci numbers is given by*

$$\sum_{k=1}^{n} F_k = F_{n+2} - 1. \tag{2.33}$$

Proof. We proceed by induction.

For $n = 1$, $\sum_{k=1}^{n} F_k = F_1 = 1$ and $F_{n+2} - 1 = F_3 - 1 = 1$. Similarly, for $n = 2$, $\sum_{k=1}^{n} F_k = F_1 + F_2 = 2$ and $F_{n+2} - 1 = F_4 - 1 = 3 - 1 = 2$. Hence the property is verified for $n = 1$ and $n = 2$ ([31]).

Assume now that $\sum_{k=1}^{n} F_k = F_{n+2} - 1$ for a given $n \geq 1$. We want to show that $\sum_{k=1}^{n+1} F_k = F_{n+3} - 1$. To this end, note that (keeping Remark 2.66 in mind)

$$\begin{aligned} \sum_{k=1}^{n+1} F_k &= \sum_{k=1}^{n} F_k + F_{n+1} \\ &= F_{n+2} - 1 + F_{n+1} \end{aligned}$$

by the inductive hypothesis. Since $F_{n+2} + F_{n+1} = F_{n+3}$ by (2.32), we conclude that

$$\sum_{k=1}^{n+1} F_k = F_{n+3} - 1,$$

so that (2.33) is true for every $n \in \mathbb{N}$ by induction. □

Exercise 2.96. Show that

$$\sum_{k=1}^{n} F_{2k-1} = F_{2n}$$

for every $n \in \mathbb{N}$.

Exercise 2.97. Show that

$$\sum_{k=1}^{n} F_k^2 = F_n \cdot F_{n+1} \tag{2.34}$$

for every $n \in \mathbb{N}$.

Proposition 2.98. *For every natural number n, the Fibonacci number F_{3n} is even.*

Proof. We proceed by induction. For $n = 1$, the number $F_3 = 2$ is even. For $n = 2$, the number $F_6 = 8$ is even.

Assume that F_{3n} is even for some n. Then ([32])

$$\begin{aligned} F_{3(n+1)} = F_{3n+3} &\overset{(2.32)}{=} F_{3n+2} + F_{3n+1} \overset{(2.32)}{=} F_{3n+1} + F_{3n} + F_{3n+1} \\ &= 2F_{3n+1} + F_{3n}. \end{aligned}$$

[31] Note that we do not really need to check the case $n = 2$, but this is safer in the context where we use the definition of F_n via (2.32), which is inductive *of depth* 2, that is, a given F_n depends on the previous 2 terms of the sequence. Sometimes in using this definition, we really assume more that $P(n)$, namely both $P(n)$ and $P(n-1)$, in deducing $P(n+1)$. In such a case, we need to include $P(1)$ **and** $P(2)$ in the base case. To be on the safe side, we'll check for the first two values of n whenever we prove something on Fibonacci numbers.

If a term depended on the previous 3 terms, then it would be safer to check $P(1)$, $P(2)$, and $P(3)$ for the base case, for you may not realize that you are assuming more than $P(n)$ to get $P(n+1)$.

[32] Note that the inductive assumption is on F_{3n}. Thus, in starting from F_{3n+3}, we use (2.32) twice to get an expression that depends on F_{3n}. It turns out that the remaining term is a multiple of 2, hence is even.

Since F_{3n} is even by inductive hypothesis, $F_{3(n+1)}$ is the sum of two even numbers, hence is even. We conclude by induction that F_{3n} is even for every natural number n. □

Exercise 2.99. Show that 5 divides F_{5n} for every natural number n.

Proposition 2.100 (Cassini's Identity). *For every natural number $n \geq 2$,*

$$F_{n-1}F_{n+1} - F_n^2 = (-1)^n. \tag{2.35}$$

Proof. We proceed by induction. For $n = 2$,

$$F_{n-1}F_{n+1} - F_n^2 = F_1F_3 - F_2^2 = 2 - 1^2 = 1 = (-1)^2.$$

Similarly, for $n = 3$,

$$F_{n-1}F_{n+1} - F_n^2 = F_2F_4 - F_3^2 = 3 - 2^2 = -1 = (-1)^3.$$

Assume now that (2.35) is true for some $n \geq 2$. We want to show that

$$F_nF_{n+2} - F_{n+1}^2 = (-1)^{n+1}.$$

To this end, note that replacing F_{n+2} by $F_{n+1} + F_n$,

$$\begin{aligned} F_nF_{n+2} - F_{n+1}^2 &= F_n(F_{n+1} + F_n) - F_{n+1}^2 \\ &= F_nF_{n+1} + F_n^2 - F_{n+1}^2 \end{aligned}$$

and

$$F_{n+1}^2 = F_{n+1}(F_n + F_{n-1}) = F_{n+1}F_n + F_{n+1}F_{n-1}$$

so that

$$F_nF_{n+2} - F_{n+1}^2 = F_nF_{n+1} + \left(F_n^2 - F_{n+1}F_{n-1}\right) - F_{n+1}F_n.$$

By inductive hypothesis, the content of the parenthesis can be replaced by $-(-1)^n = (-1)^{n+1}$. Moreover, the other two terms cancel out, to the effect that

$$F_nF_{n+2} - F_{n+1}^2 = (-1)^{n+1}.$$

We conclude by induction that the identity is true for all $n \geq 2$. □

Corollary 2.101. *For every natural number n,*

$$F_{n+1}^2 - F_{n+1}F_n - F_n^2 = (-1)^n. \tag{2.36}$$

Proof. Since

$$F_{n+1}^2 - F_{n+1}F_n = F_{n+1}\left(F_{n+1} - F_n\right) \overset{(2.32)}{=} F_{n+1}F_{n-1},$$

this follows immediately from Proposition 2.100. □

Let us consider what (2.36) tells us about the *asymptotic behavior* (that is, the behavior for large n) of the sequence $\{F_n\}_{n=1}^{\infty}$. Dividing both sides of (2.36) by the positive number F_n^2 yields

$$\left(\frac{F_{n+1}}{F_n}\right)^2 - \frac{F_{n+1}}{F_n} - 1 = \frac{(-1)^n}{F_n^2}.$$

Since F_n grows without bounds, the right-hand side $\frac{(-1)^n}{F_n^2}$ quickly becomes very close to 0 as n grows. Hence, as n grows, the ratio of consecutive terms

$$\frac{F_{n+1}}{F_n}$$

gets closer and closer to a root of the quadratic equation

$$x^2 - x - 1 = 0,$$

whose solutions are $\frac{1+\sqrt{5}}{2}$ and $\frac{1-\sqrt{5}}{2}$. Since Fibonacci numbers are positive, we conclude that

$$\lim_{n\to\infty} \frac{F_{n+1}}{F_n} = \frac{1+\sqrt{5}}{2}.$$

In other words, the sequence $\{F_n\}_{n=1}^{\infty}$ asymptotically approaches a geometric sequence of common ratio the *golden ratio*

$$\Phi = \frac{1+\sqrt{5}}{2}.$$

This observation can help us obtain an exact formula for the nth Fibonacci number. Note that as n grows, F_n gets closer and closer to the terms of a geometric sequence of the form $c\Phi^n$ for some constant c but does not exactly match such a sequence. Note also that

$$\Phi_* = 1 - \Phi = \frac{1-\sqrt{5}}{2}$$

is the other root of

$$x^2 - x - 1 = 0, \tag{2.37}$$

and that therefore the difference

$$y_n = c\Phi^n - c\Phi_*^n$$

satisfies the same inductive relation

$$y_{n+2} = y_{n+1} + y_n \tag{2.38}$$

as the Fibonacci numbers. Indeed,

$$\begin{aligned} y_{n+1}+y_n &= c\left(\Phi^{n+1}+\Phi^n-\left(\Phi_*^{n+1}+\Phi_*^n\right)\right) \\ &= c\left(\Phi^n(\Phi+1)-\Phi_*^n(\Phi_*+1)\right) \\ &= c\left(\Phi^n\cdot\Phi^2-\Phi_*^n\cdot\Phi_*^2\right) \end{aligned}$$

because Φ and Φ_* satisfy (2.37), that is, satisfy $1+x=x^2$. Thus we have verified (2.38). That means that the sequence $\{y_n\}_{n=1}^{\infty}$ would match the sequence $\{F_n\}_{n=1}^{\infty}$ if we could find a constant c such that $y_1=F_1$ and $y_2=F_2$. Let's try!

$$y_1 = c\left(\frac{1+\sqrt{5}}{2}-\frac{1-\sqrt{5}}{2}\right) = c\sqrt{5}$$

and

$$\begin{aligned} y_2 &= c\left(\left(\frac{1+\sqrt{5}}{2}\right)^2-\left(\frac{1-\sqrt{5}}{2}\right)^2\right) \\ &= \frac{c}{4}\left(1+2\sqrt{5}+5-(1-2\sqrt{5}+5)\right) = c\sqrt{5}, \end{aligned}$$

so that taking $c=\frac{1}{\sqrt{5}}$ yields $y_1=F_1$ and $y_2=F_2$ as desired. Thus

Theorem 2.102. *For every natural number n, the nth Fibonacci number is*

$$F_n = \frac{1}{\sqrt{5}}\left(\left(\frac{1+\sqrt{5}}{2}\right)^n-\left(\frac{1-\sqrt{5}}{2}\right)^n\right).$$

The considerations preceding Theorem 2.102 are *essentially a proof*, and at any rate are important to understand where the result comes from. But they are not quite a proper proof, which you should try to write out for yourself.

We conclude this section with a proof by induction that is slightly different from those we have seen so far. We want to show a pair of properties of $\{F_n\}_{n=1}^{\infty}$ and we are going to prove *both at once* by induction, because they are interdependent.

Theorem 2.103. *For every natural number $n\geq 2$, we have*

$$F_{2n-1} = F_n^2+F_{n-1}^2$$

and

$$F_{2n} = F_{n+1}F_n+F_nF_{n-1}.$$

Proof. We prove both formulas by induction on n, that is, $P(n)$ is

$$F_{2n-1} = F_n^2+F_{n-1}^2 \textbf{ and } F_{2n} = F_{n+1}F_n+F_nF_{n-1}. \tag{2.39}$$

When $n=2$, $F_{2n-1}=F_3=2$ and $F_n^2+F_{n-1}^2=1+1=2$. On the other hand, $F_{2n}=F_4=3$ and $F_{n+1}F_n+F_nF_{n-1}=2+1=3$.

When $n=3$, $F_{2n-1}=F_5=5$ and $F_n^2+F_{n-1}^2=2^2+1^2=5$. On the other hand, $F_{2n}=F_6=8$ and $F_{n+1}F_n+F_nF_{n-1}=3\times 2+2\times 1=8$.

Assume now that (2.39) is true for some $n\geq 2$.

Since

$$\begin{aligned}
F_{n+1}^2+F_n^2 &\overset{(2.32)}{=} (F_n+F_{n-1})^2+F_n^2\\
&= F_n^2+F_{n-1}^2+F_n^2+2F_nF_{n-1}\\
&= \left(F_n^2+F_{n-1}^2\right)+\left(F_n^2+F_nF_{n-1}\right)+F_nF_{n-1}\\
&= \left(F_n^2+F_{n-1}^2\right)+F_n\left(F_n+F_{n-1}\right)+F_nF_{n-1}\\
&= \left(F_n^2+F_{n-1}^2\right)+F_nF_{n+1}+F_nF_{n-1}
\end{aligned}$$

and by inductive assumption $F_n^2+F_{n-1}^2=F_{2n-1}$ and $F_nF_{n+1}+F_nF_{n-1}=F_{2n}$, we obtain

$$F_{n+1}^2+F_n^2=F_{2n-1}+F_{2n}\overset{(2.32)}{=}F_{2n+1}.$$

On the other hand,

$$\begin{aligned}
F_{n+2}F_{n+1}+F_{n+1}F_n &\overset{(2.32)}{=} (F_{n+1}+F_n)\,F_{n+1}+(F_n+F_{n-1})\,F_n\\
&= F_{n+1}^2+F_n^2+F_nF_{n+1}+F_nF_{n-1}.
\end{aligned}$$

We have just proved that $F_{n+1}^2+F_n^2=F_{2n+1}$ and moreover $F_nF_{n+1}+F_nF_{n-1}=F_{2n}$ by inductive hypothesis. Hence,

$$F_{n+2}F_{n+1}+F_{n+1}F_n=F_{2n+1}+F_{2n}\overset{(2.32)}{=}F_{2n+2}.$$

We conclude by induction that (2.39) is true for all $n\geq 2$. □

Additional Exercises (Homework)

Exercise 2.104. Show that for every $n\in\mathbb{N}$,

$$\sum_{i=1}^{n} i^3=\frac{n^2(n+1)^2}{4}.$$

Exercise 2.105. Show that for every $n\in\mathbb{N}$,

$$\sum_{i=1}^{n} i(i+1)=\frac{n(n+1)(n+2)}{3}.$$

Exercise 2.106. Show that for every $n\in\mathbb{N}$,

$$\sum_{i=1}^{n}(3i-2)=\frac{n(3n-1)}{2}.$$

Exercise 2.107. Show that for every $n \in \mathbb{N}$,

$$\sum_{i=1}^{n}(2i-1)^3 = n^2(2n^2-1).$$

Exercise 2.108. Show that for every $n \in \mathbb{N}$,

$$\sum_{i=1}^{n}\frac{i}{(i+1)!} = 1 - \frac{1}{(n+1)!}.$$

Exercise 2.109. Show that $3|(n^3+5n+6)$ for every integer $n \geq 0$.

Exercise 2.110. Show that $9|(4^{3n}+8)$ for every integer $n \geq 0$.

Exercise 2.111. Show that

$$2^n + 1 \leq 3^n$$

for every $n \in \mathbb{N}$.

Exercise 2.112. Prove that

$$\sum_{i=1}^{n}\frac{1}{i^2} \leq 2 - \frac{1}{n}$$

for all $n \in \mathbb{N}$.

Exercise 2.113. Show that

$$\prod_{i=1}^{n}\frac{1}{i} \leq \frac{1}{2^n}$$

for all integer $n \geq 4$.

Exercise 2.114. Rewrite the proof of Proposition 1.112 using induction formally.

Exercise 2.115. Show that

$$\prod_{i=1}^{n}(4i-2) = \frac{(2n)!}{n!}$$

for all $n \in \mathbb{N}$.

Exercise 2.116. Show that for every $n \in \mathbb{N}$,

$$\frac{n^3}{3} + \frac{n^5}{5} + \frac{7n}{15}$$

is an integer.

Exercise 2.117. Prove Theorem 1.120 by induction.

Exercise 2.118. Show that

$$\frac{F_{2n+2}}{F_{n+1}} = \frac{F_{2n}}{F_n} + \frac{F_{2n-2}}{F_{n-1}}$$

for every $n \geq 2$ and deduce that F_n divides F_{2n} for every $n \geq 1$.

2.7 A Word on Style

Rather than reformulate all of it, I would recommend reading Section 5.3 of [19] on Mathematical Writing (freely available online) that provides valuable pieces of advice and examples of do's and don'ts. Let me however summarize the most important points. The following conventions help make things more readable and rule out some ambiguities:

- Never use symbols as "shortcuts." For instance, saying "these two sets are =" is comparable to use "4u" to say "for you." This has no place in formal writing. Similarly, quantifiers and logical connectives ($\forall, \exists, \vee, \wedge, \Longrightarrow, \Longleftrightarrow$) can only be used in displayed formulas, but not as part of in line text. In the text, spell out "for every," "there is," "if...then," "if and only if," and so on.
- Recall also that quantifiers precede the formula to which they apply, as discussed in Section 1.3.3.
- avoid starting a sentence with a formula or symbol. For instance, in a text, rather than

  $$A \text{ is a subset of } B$$

  you may say

  $$\text{The set } A \text{ is a subset of } B.$$

- avoid placing two formulas on either side of a comma. For instance, rather than

  $$\text{As } x^2 - 9 = 0,\ x = \pm 3$$

  you may write

  $$\text{As } x^2 - 9 = 0, \text{ it follows that } x = \pm 3.$$

- A displayed formula may be the end of a sentence, in which case it should end with a period, or may be part of a sentence that continues afterwards, in which case it should end with a comma.
- "it" is your enemy! Watch out whenever you use "it." Make sure that it is crystal clear what "it" refers to in your sentence. If there is any room for ambiguity, spell out what "it" was supposed to refer to.

2.8 Typesetting Mathematics

Typesetting mathematics with a standard word processor such as MS Word is very cumbersome. On the other hand, LaTeX is a powerful programming language for scientific typesetting and creates high quality output. Installing and using LaTeX is free. A wikipedia search for Latex will get you a wealth of information on how to get started if you have a lot of time to invest to learn LaTeX from scratch. A good compromise is the free LaTeX front-end called LyX.

LyX is a powerful free *what you see is what you mean* word processor. In other words, like LaTeX, there is a compiling process to generate the output. However, what you see is not code, but something close to what you mean to obtain as an output, albeit with a different formatting. The present document was produced with LyX. The spirit is that you should not be concerned with formatting, which will be handled automatically by the software. This way you can focus on content. All formatting is handled automatically by placing most elements inside *environments*, such as Definition, Theorem, Proof, itemized list, enumeration, etc. Each environment has its own formatting. There are many more advantages to using LyX including a powerful cross-referencing system, and easy to generate lists (index, bibliography, list of symbols, table of contents, etc.).

To get started with LyX you can start on my webpage [23] where you will find links to written tutorials, and some video tutorials.

Suggested Further Readings

Many examples in this chapter follow [19], where you can find a few more interesting examples.

The best way to assimilate proof techniques and to learn how to write proofs properly is to read well-written proofs that are accessible to you. From now on, pay particular attention when something is proved in one of your classes.

If you find the exposition of proof techniques here too dense, you may enjoy [28], a book that breaks down various types of proof arguments and goes through a large list of elementary examples with extensive comments—a little too much for my taste, but you may view it differently. At any rate, it contains an extensive repertoire of commented step by step elementary proofs. The companion website also has some video lectures to accompany the book.

Complements on induction, particularly the equivalence of the induction principle, strong induction principle, and well-ordering principle are to be found in the Appendix.

Chapter 3
Relations

Why More Formalism?

More than on proofs, this is a text on foundations of mathematics, which started with examining why we may need a formal language to do mathematics. Our first chapter developed enough formalism for us to practice standard proof techniques in the second chapter. We now return to the development of standard foundational formalism, because these conventions and concepts are everywhere in higher level mathematics, and are part of the core vocabulary of mathematicians.

Relations provide a broad and flexible framework to formalize the notion of two objects being related, even though this is not in a quantified manner: two things are related, or they are not. This is a very general concept, and thus a ubiquitous one, which is why we will examine the general notion and related concepts.

Yet, we are most interested in very special relations: *functional* relations, *equivalence* relations, and *order* relations. Functional relations define functions, which makes it plain why they are important. Equivalence relations and the companion notion of quotient set will seem abstract at first, but is nothing but an abstract model for a fundamental thought process: that of identifying similitudes, grouping objects accordingly, and "forgetting" their differences to form concepts (see Remark 3.91 for a more thorough explanation). Because what it models is at the heart of how our brains process experience, it comes as no surprise that introducing equivalence relations and passing to the quotient set is a fundamental and ubiquitous operation in mathematics. Hence we will take the time to examine several points of view on the same process.

Similarly, order relations model our ability to make judgments of value such as "this is better than that," where "better" is always from a certain point of view, a certain set of criteria, that is, a certain *order relation* (see Remark 3.48 for a more thorough explanation). This process of evaluation is a fundamental part of how we

F. Mynard, *An Introduction to the Language of Mathematics*,
https://doi.org/10.1007/978-3-030-00641-9_3

try to order and thus understand the world. Expectedly, many aspects of mathematics can be understood from the point of view of order theory. We thus will take the time to introduce some of the most common and useful concepts related to order relations.

3.1 General Relations

Definition 3.1. A *relation* R from X to Y is a subset of $X \times Y$. Traditionally, we write

$$xRy$$

to denote that $(x,y) \in R$, that is, that x is R*-related to* y, and

$$x \not R y$$

to denote $(x,y) \notin R$, that is, that x is not R-related to y.

Remark 3.2. Note that at this level of generality, in checking if xRy for a given $x \in X$ and $y \in Y$, we examine whether an object in X is related to one in a *potentially different set* Y. However, we will soon focus on the case $X = Y$, thus examining whether two elements of the same set are related or not from the viewpoint of a relation R. Even in this case, you should be mindful of the fact that a relation does not need to be symmetric: we may have xRy but $y \not R x$.

Definition 3.3. For every $x \in X$, let

$$R(x) = R(\{x\}) = \{y \in Y : (x,y) \in R\}$$

denote the set of elements of Y that are R-related to x, and, if $A \subset X$, we define

$$R(A) = \bigcup_{x \in A} R(x) = \{y \in Y : \exists x \in A \; (x,y) \in R\}.$$

Remark 3.4. Note that $R(x)$ is a shorthand for the *set* $R(\{x\})$. Hence $R(x)$ is not an element of Y (as it would if R were a function $R : X \to Y$), but a subset of Y. Note also that this subset might be empty! In introducing *the image* $R(x)$ of $x \in X$ under a relation R, we think of a relation as a generalized function, which may be multi-valued, and may take $\emptyset$ as a value. Hence you may encounter sources calling our relations *multi-valued maps.*

Definition 3.5. The *inverse relation* R^{-1} of R is the relation

$$R^{-1} = \{(y,x) \in Y \times X : (x,y) \in R\}$$

from Y to X. Of course

$$x \in R^{-1}(y) \iff y \in R(x) \iff (x,y) \in R.$$

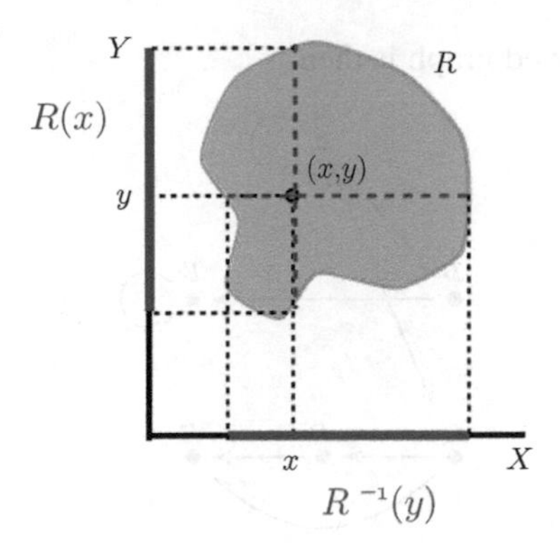

Exercise 3.6. Let R and S be relations on X. Show that

$$R \subset S \iff R^{-1} \subset S^{-1}.$$

Example 3.7. Let X be a set of people, say

$$X = \{\text{Bob, Anna, Max, Maria}\}$$

and Y be a set of types of animal, say

$$Y = \{\text{cat, dog, goldfish, snake, parrot}\}.$$

Define the relation R by xRy if x has y as a pet. If, for instance,

$$R=\{(\text{Bob,cat}),(\text{Bob,dog}),(\text{Bob,snake}),(\text{Anna,cat}),(\text{Anna,glodfish}),(\text{Max,dog})\},$$

then $R(\text{Bob}) = \{\text{cat,dog,snake}\}$, which means that Bob has a cat, a dog, and a snake. On the other hand, $R^{-1}(\text{dog}) = \{\text{Bob,Max}\}$ which means that people in X who have a dog are Bob and Max. Note also that $R(\text{Maria}) = \emptyset$ which means that Maria doesn't have pets, and $R^{-1}(\text{parrot}) = \emptyset$, which means that none of the people in X have a parrot.

Many relations relate objects of the same set. We say that R is a *relation on* X if $R \subset X \times X$.

Remark 3.8. In the case of a finite set X, we can represent the relation as a *directed graph*, that is, we consider each point of X as a vertex, and we draw a directed edge from x to y if and only if xRy. If xRx, then we obtain a loop at x. In this representation, R^{-1} is represented by the graph obtained by reversing all the arrows. Of course, conversely, a directed graph can be understood as a relation on the set of its vertices.

Example 3.9. Let $X = \{A,B,C,D,E,F\}$ with $R \subset X \times X$ given by

$$R = \{(A,A),(A,B),(A,C),(B,F)(C,D),(C,E),(D,E),(F,F)\}.$$

The corresponding directed graph is then

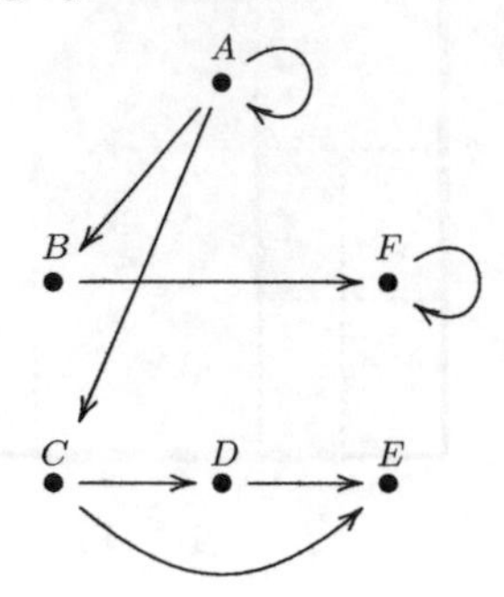

and thus, that of R^{-1} is

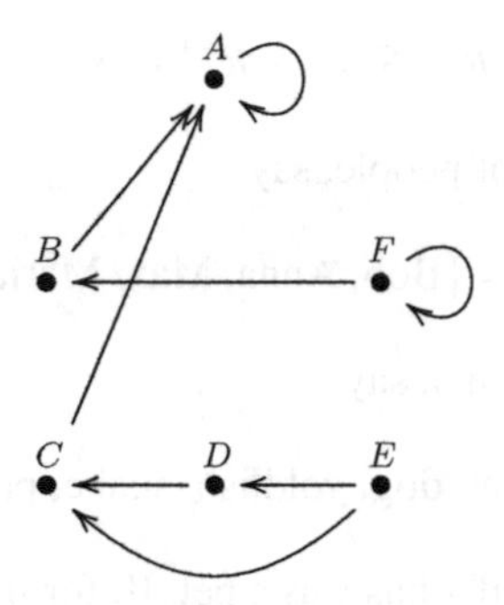

Definition 3.10. A relation R on X (that is, $R \subset X \times X$) is:

- *reflexive* if xRx for all $x \in X$;
- *symmetric* if for all $x, y \in X$

$$xRy \Longrightarrow yRx; \tag{3.1}$$

- *antisymmetric* if for all $x, y \in X$

$$xRy \text{ and } yRx \Longrightarrow x = y; \tag{3.2}$$

- *transitive* if for all $x, y, z \in X$

$$xRy \text{ and } yRz \Longrightarrow xRz. \tag{3.3}$$

- an *equivalence relation* if it is reflexive, symmetric, and transitive;
- a (partial) *order* if it is reflexive, antisymmetric, and transitive.

Let us start with simple reinterpretations of the definitions. To this end, first define:

Definition 3.11. The *identity relation on X* is

$$I_X = \{(x,x) : x \in X\}.$$

We will see that the identity relation plays a particular role. Note that I_X is reflexive, transitive, and symmetric, hence an equivalence relation (the trivial equivalence!), and it is also antisymmetric, hence it is an order relation too (the trivial order)!

Proposition 3.12. *A relation R on X is:*

1. *reflexive if and only if $I_X \subset R$ if and only if each vertex of its directed graph carries a loop :*

2. *symmetric if and only if $R = R^{-1}$ if and only if each directed edge goes in both direction:*

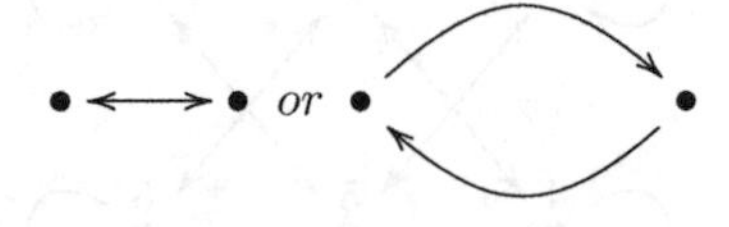

3. *transitive if and only if in every situation like below in the directed graph, the dotted arrow exists*

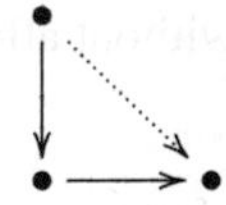

4. *antisymmetric if and only if $R \cap R^{-1} \subset I_X$, if and only if the directed graph has no two directional edges but loops.*

Proof. The graph theoretic interpretations are clear. As for the set theoretic ones: (1) R is reflexive if and only if $(x,x) \in R$ for all $x \in X$, that is, if $I_X \subset R$. (2) R is symmetric if and only if

$$xRy \Longleftrightarrow yRx$$

for all $x, y \in X$, that is, if $(x,y) \in R$ if and only if $(y,x) \in R$. Since $(x,y) \in R$ if and only if $(y,x) \in R^{-1}$, we conclude that R is symmetric if and only if $R = R^{-1}$.

(4) xRy and yRx if and only if $(x,y) \in R \cap R^{-1}$, and $x = y$ if and only if $(x,y) \in I_X$. Hence

$$xRy \text{ and } yRx \Longrightarrow x = y$$

is equivalent to $R \cap R^{-1} \subset I_X$. □

Corollary 3.13. *If a relation R on a set X is both symmetric and antisymmetric, then $R \subset I_X$. If moreover R is reflexive, then $R = I_X$.*

Proof. If R is both symmetric and antisymmetric, then by Proposition 3.12 (2) and (4), $R = R^{-1}$ and $R \cap R^{-1} \subset I_X$ so that $R \subset I_X$. If moreover R is reflexive, then $I_X \subset R$ by Proposition 3.12 (1). Hence $R = I_X$. □

Example 3.14. Let R be a relation on $X = \{1,2,3,4,5,6\}$ defined by xRy if and only if x and y are either both odd or both even, that is, x and y have the same parity. In other words,

$$R = \{(1,1),(1,3),(3,1),(1,5),(5,1),(3,3),(3,5),(5,3),(5,5),\\ (2,2),(2,4),(4,2),(2,6),(6,2),(4,4),(4,6),(6,4),(6,6)\}.$$

For instance, $R(2) = \{2,4,6\} = R(4) = R(6)$ and $R(1) = \{1,3,5\} = R(3) = R(5)$. Since R is *symmetric*, that is, xRy if and only if yRx, we have $R = R^{-1}$. Here xRx for all $x \in X$, that is, R is *reflexive*. If xRy and yRz, then xRz, that is R is *transitive*. Hence R is an *equivalence relation*. The corresponding graph is

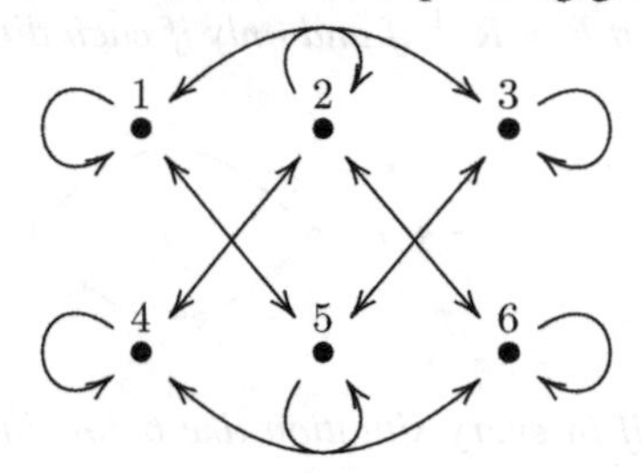

As noted in Proposition 3.12, the fact that $R = R^{-1}$ visually corresponds to the fact that all arrows can be reversed without affecting the graph.

Example 3.15. Let $R \subset \mathbb{R} \times \mathbb{R}$ be

$$R = \{(x,y) : x \leq y\}.$$

Then R is a relation on $\mathbb{R}$ defined by

$$xRy \iff x \leq y,$$

that is, R is the relation "less than or equal to." Note that R is reflexive, because $x \leq x$ for all $x \in \mathbb{R}$ and transitive because if $x \leq y \leq z$ then $x \leq z$, but R is not symmetric: for instance, $1 \leq 2$ but $2 \not\leq 1$. On the other hand, it is antisymmetric: if $x \leq y$ and $y \leq x$, then $x = y$. Hence R is an *order relation*. It is, of course, the order relation you have in mind when we say "order." But we will see that there are order relations that are quite different from this "standard" order.

Exercise 3.16. Which of the properties of Definition 3.10 are satisfied in the following examples:

1. The relation

$$R = \{(a,a),(a,b),(b,a),(b,b),(b,c),(c,b),(a,c)\}$$

   on $X = \{a,b,c\}$.
2. The relation R on $\mathbb{Z}$ where xRy if and only if x and y have at least a digit in common in their decimal expansions.
3. The relation $<$ on $\mathbb{Z}$;
4. the relation $\leq$ on $\mathbb{Z}$;

5. the relation $\subset$ on $\mathbb{P}X$;
6. the relation $=$ on $\mathbb{Z}$;
7. The relation R on a set X of people defined by xRy if and only if x and y have the same height (rounding to the nearest centimeter);
8. Let X be a set of finite sets and let R be a relation on X defined by $(A,B) \in R$ if and only if A and B have the same number of elements.
9. the relation $|$ defined on $\mathbb{Z}$ by $a|b$ if there is $k \in \mathbb{Z}$ with $b = ka$;
10. the relation $|$ on $\mathbb{N}$.

Exercise 3.17. Show that if R is a relation on X and

1. R is reflexive, then so is R^{-1};
2. R is transitive, then so is R^{-1};
3. R is symmetric, then so is R^{-1};
4. R is antisymmetric, then so is R^{-1}.

In particular, if R is an order relation, then so is R^{-1}, which is then called the *reverse order* of R.

Example 3.18. The reverse order of the order relation $\leq$ on $\mathbb{R}$ (of Example 3.15) is the relation $\geq$ on $\mathbb{R}$. The reverse order of the order relation $\subset$ on $\mathbb{P}X$ (of Exercise 3.16 (5)) is $\supset$. The reverse order of the relation "divides" on $\mathbb{N}$ (of Exercise 3.16 (10)) is the relation aRb if there is $k \in \mathbb{N}$ with $a = kb$, that is, R is the relation "is a multiple of."

Example 3.19. Let $A = \{a,b,c,d\}$ and $R \subset A \times A$ be

$$R = \{(a,b),(b,c),(c,c),(d,c)\}.$$

For instance, $R(a) = \{b\}$ and $R^{-1}(c) = \{c,b,d\}$. The corresponding directed graph is

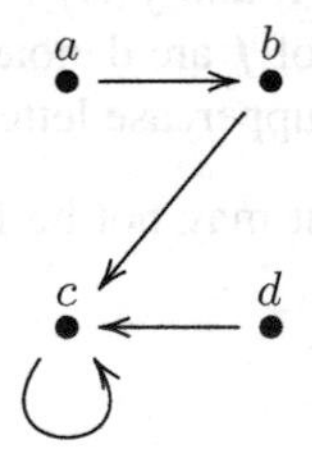

Since $a \not R\, a$, the relation R is not reflexive. On the other hand, aRb but $b \not R\, a$, that is, R is not symmetric. Moreover, aRb and bRc but $a \not R\, c$, that is, R is not transitive. However, this is a *functional relation*, in the sense that this is the graph of the function $r : A \to A$ defined by $r(a) = b$, $r(b) = r(c) = r(d) = c$ (see Definition 3.21 below). This can be detected on the directed graph by noting that only one arrow is leaving from each vertex.

Let us turn now more formally to functional relation. First note:

Example 3.20. The graph of a function $f : X \to Y$ is a subset of $X \times Y$ and can thus be seen as a relation $\tilde{f}$. Note that

$$y \in \tilde{f}(x) \iff (x,y) \in \tilde{f} \iff y = f(x).$$

For the graph of a function, the set $\tilde{f}(x)$ is always a *singleton*, that is, has a single element.

3.1.1 Functional Relations

Definition 3.21. A relation $R \subset X \times Y$ is *functional* if for every $x \in X$, the set $R(x)$ is a singleton.

In view of Remark 1.135, a relation is functional if and only if it is a function $r : X \to Y$ (seen as a subset of $X \times Y$, that is, once identified with its graph), where $r(x)$ is the unique member of $R(x)$.

Remark 3.22. The distinction between a function and a functional relation (between a function f and its graph $\tilde{f}$) is artificial and unnecessary (as pointed out in Remark 1.135, the proper way to define a function is as a functional relation!), but helps us distinguish functions among relations, even though we first introduced functions through a different formalism, long before we introduced relations.

Note that $\tilde{f}(A) = f[A]$ and $\tilde{f}(x) = \{f(x)\}$, that is, $y \in \tilde{f}(x)$ if and only if $y = f(x)$. Hence there is no need to take pain in distinguishing $\tilde{f}$ and f, and thus $\tilde{f}(A)$ from $f(A)$ from $f[A]$.

Oftentimes, the reader is expected to rule out ambiguities from context—for instance, the distinction between $f[A]$ and $f(A)$ is not necessary if context makes it clear that elements of the domain of f are denoted by lowercase letters, while subsets of the domain are denoted by uppercase letters.

Of course, $\tilde{f}^{-1}$ is a relation, but may not be functional unless f is bijective, as we will see in Exercise 3.24.

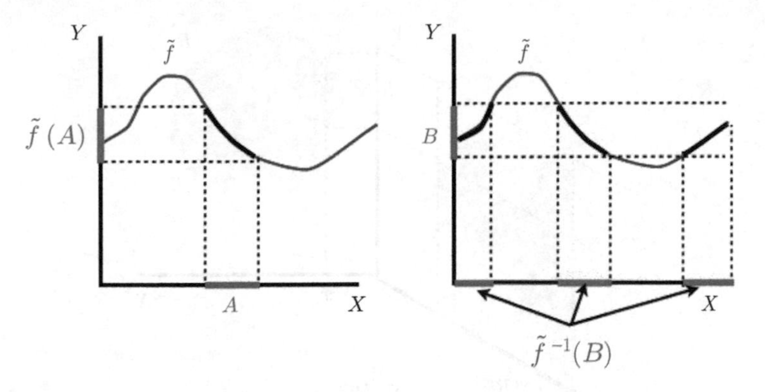

Definition 3.23. A relation $R \subset X \times Y$ is *surjective* if $R(X) = Y$ and *injective* if

$$x_1 \neq x_2 \Longrightarrow R(x_1) \cap R(x_2) = \emptyset.$$

Exercise 3.24. Let $R \subset X \times Y$. Show that:

1. a relation R is functional if and only if R^{-1} is injective and surjective.
2. A functional relation is injective (respectively, surjective) if and only if the corresponding function is.
3. If $f : X \to Y$ is one-to-one, then $\tilde{f}^{-1}$ is a functional relation from $f[X]$ to X. ([1])

Definition 3.25 (Composition of Relations). If $R \subset X \times Y$ and $S \subset Y \times Z$ then the *composite* $S \circ R$ is defined as

$$S \circ R = \{(x,z) : \exists y \in Y \; (x,y) \in R \text{ and } (y,z) \in S\}. \tag{3.4}$$

In other words, $x(S \circ R)z$ if and only if there is $y \in Y$ with xRy and ySz, equivalently

$$x(S \circ R)z \iff R(x) \cap S^{-1}(z) \neq \emptyset. \tag{3.5}$$

[1]Note that the function $g : X \to f[X]$ defined by $f(x) = g(x)$ for all $x \in X$ is a bijection and its inverse in the sense of Definition 1.110 is the function induced by the functional relation $\tilde{f}^{-1}$.

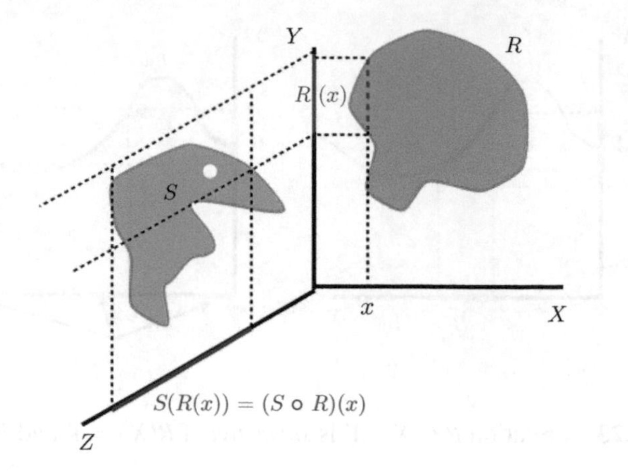

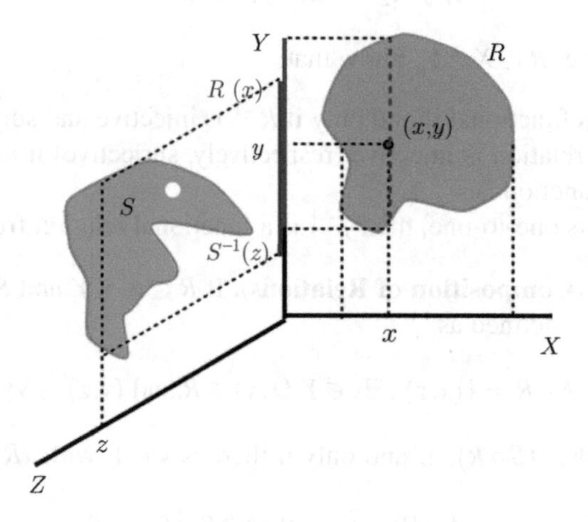

Exercise 3.26. Show that if $R \subset X \times Y$ and $S \subset Y \times Z$ are both functional relations (with induced functions r and s) then $S \circ R$ is the graph of the function $s \circ r : X \to Z$, and is thus functional.

Recall that I_X denotes the identity relation on the set X. It is a functional relation: it is the graph of the identity function of X.

Exercise 3.27. Let $R \subset X \times Y$ and $S \subset Y \times Z$. Show that:

1. for every $A \subset X$,
$$(S \circ R)(A) = S(R(A)).$$

2.
$$(S \circ R)^{-1} = R^{-1} \circ S^{-1}.$$

3. Show that if $R \subset X \times Y$ then $R \circ I_X = I_Y \circ R = R$.

Exercise 3.28. Show that a relation R on a set X is transitive if and only if

$$R \circ R \subset R.$$

Additional Exercises (Homework)

Exercise 3.29. Let
$$X = \{1,3,5,7,15,21,30\}.$$

1. Write out the "divides" relation $|$ on X (as a subset of $X \times X$) and draw the corresponding graph.
2. Write out the "less than" relation $<$ on X (as a subset of $X \times X$) and draw the corresponding graph.
3. Write out the "greater than or equal" relation $\geq$ on X (as a subset of $X \times X$) and draw the corresponding graph.

Exercise 3.30. Let X be a non-empty set and let

$$R = (X \times X) \setminus I_X.$$

What usual symbol can be used instead of R in writing xRt?

Exercise 3.31. Consider the relation on $X = \{a,b,c,d\}$ given by the graph

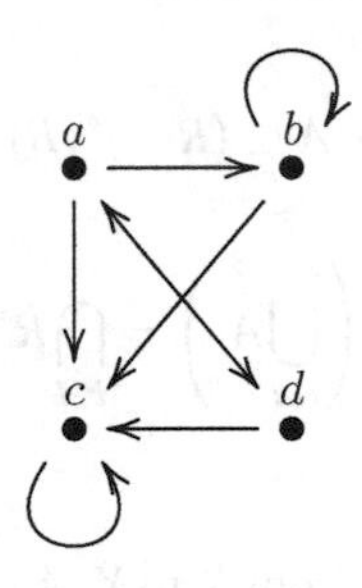

1. Write this relation as a subset of $X \times X$.
2. Which of the properties of Definition 3.10 does it satisfy? Explain.
3. Is this relation functional?

Exercise 3.32. Which of the properties of Definition 3.10 do the following relations satisfy? Explain.

1. The relation R on $\mathbb{Z}$ defined by xRy if and only if $x^2 = y^2$;
2. The relation $R = \{(1,3),(1,1),(2,4),(5,4),(4,2),(5,5)\}$ on $X = \{1,2,3,4,5\}$;
3. Th relation $R = \{(1,2),(2,3),(1,3),(1,1)\}$ on $X = \{1,2,3\}$;
4. Th relation $R = \{(1,2),(2,3),(1,3),(1,1),(3,1)\}$ on $X = \{1,2,3\}$.

Exercise 3.33. Consider the empty relation $R = \emptyset$ on a non-empty set X. Which of the properties of Definition 3.10 does it satisfy? Explain.

Exercise 3.34. How many relations are there on a finite set X with n elements? Explain.

Exercise 3.35. Describe with a graph each one of the relations on $X=\{a,b\}$.

Exercise 3.36. Consider the four properties reflexive (R), symmetric (S), antisymmetric (A), and transitive (T). There are $2^4=16$ possible combinations of these properties. Find an example of relation illustrating each possible combination.

Exercise 3.37. Consider the relation R on $\mathbb{R}$ defined by xRy if $|x-y|<1$.

1. Which of the properties of Definition 3.10 does it satisfy? Explain.
2. Which of the properties of Definition 3.10 does the same relation defined on $\mathbb{Z}$ satisfy? Explain.

Exercise 3.38. Let R be a relation on X that is both symmetric and transitive. Show that if there is $x_0 \in \bigcap_{x\in X} R^{-1}(x)$, then R is also reflexive. Describe this relation.

Exercise 3.39. Let $R\subset X\times Y$ be a relation. For each $A\subset X$, let

$$R^*(A)=\bigcap_{a\in A} R(x)$$

denote the *polar of A with respect to R*. Show

1. $$B\subset R^*(A) \iff A\subset (R^{-1})^*(B) \iff A\times B\subset R.$$
2. $$R^*\left(\bigcup_{i\in I} A_i\right)=\bigcap_{i\in I} R^*(A_i)$$
   for every $\{A_i : i\in I\}\subset \mathbb{P}X$.

Exercise 3.40. Let R be a relation from X to Y, $A\subset X$ and $B\subset Y$. Show that

$$R(A)\cap B\neq\emptyset \iff A\cap R^{-1}(B)\neq\emptyset \iff (A\times B)\cap R\neq\emptyset.$$

Exercise 3.41. Let X be a non-empty set and consider the relation # on $\mathbb{P}(\mathbb{P}X)$ defined by

$$\mathscr{A}\#\mathscr{B} \iff \forall A\in\mathscr{A}\ \forall B\in\mathscr{B}\ (A\cap B\neq\emptyset).$$

If $\mathscr{A}\#\mathscr{B}$ we say that $\mathscr{A}$ and $\mathscr{B}$ *mesh*.

1. Which of the properties of Definition 3.10 are satisfied by #?
2. Show that if $R\subset X\times Y$, $\mathscr{A}\in\mathbb{P}(\mathbb{P}X)$ and $\mathscr{B}\in\mathbb{P}(\mathbb{P}Y)$ and
   $$\begin{aligned}\mathscr{A}\times\mathscr{B}&=\{A\times B:\ A\in\mathscr{A}, B\in\mathscr{B}\}\\ R[\mathscr{A}]&=\{R(A): A\in\mathscr{A}\}\\ R^{-1}[\mathscr{B}]&=\{R^{-1}(B): B\in\mathscr{B}\}\end{aligned}$$
   then
   $$R[\mathscr{A}]\#\mathscr{B} \iff \mathscr{A}\#R^{-1}[\mathscr{B}] \iff (\mathscr{A}\times\mathscr{B})\#\{R\},$$

where the first # symbol is the meshing relation on $\mathbb{P}(\mathbb{P}Y)$, the second on $\mathbb{P}(\mathbb{P}X)$, and the third on $\mathbb{P}(\mathbb{P}(X \times Y))$.

Exercise 3.42. Let $R \subset X \times Y$ where X and Y are finite sets. For the sake of argument, let $X = \{a,b,c\}$ and $Y = \{\alpha,\beta,\gamma,\delta\}$. With the relation R, we can associate a *relational matrix* M_R which is a 3×4 matrix with entries 0 or 1. The rows can be thought of as labeled by a, b, and c, and the columns by α, β, γ, and δ. If $t \in X$ and $\phi \in Y$, we enter a 1 in the place labeled by t and ϕ if $tR\phi$, and a 0 otherwise.

1. If
$$R = \{(a,\alpha),(a,\delta),(b,\gamma),(b,\beta),(b,\delta),(c,\alpha)\},$$
give M_R.
2. Assume that X, Y, and Z are finite sets, that $R \subset X \times Y$ and $S \subset Y \times Z$, so that $S \circ R \subset X \times Z$. With the convention that entries of the relational matrices are valued in $\{0,1\}$ equipped with the operation $+$ defined by
$$0+0=0 \text{ and } 0+1=1+0=1+1=1,$$
show that
$$M_{S \circ R} = M_R \times M_S,$$
with the usual matrix multiplication (with this modified addition of entries).

3.2 Order Relations

Recall that a relation R on X that is reflexive, antisymmetric, and transitive is an *order relation*. If moreover every two elements of X are *comparable,* that is,
$$\forall x,y \in X \ (xRy \text{ or } yRx),$$
the order is called *total* or *linear*. To emphasize that an order relation does not need to be a total order, we often say a *partial order* for an order relation. A set equipped with a partial order is called a *partially ordered set* or *poset* for short.

Example 3.43. The standard order $\leq$ on $\mathbb{R}$ is total (given two numbers, one is greater or equal than the other). In contrast, the inclusion order $\subset$ on $\mathbb{P}X$ and the division order $|$ on $\mathbb{N}$ are not ([2]). For instance, if $X = \{a,b,c\}$, then $A = \{a,b\}$ and $B = \{b,c\}$ are two elements of $\mathbb{P}X$ that are not comparable under $\subset$ (that is, $A \not\subset B$ and $B \not\subset A$). Similarly, 2 and 5 are two elements of $\mathbb{N}$ that are not comparable under $|$ for 2 does not divide 5 and 5 does not divide 2.

As a **notational convention**, we use $\leq$ for a generic partial order and not only for the usual order on $\mathbb{R}$ and its subsets. If we need more than one partial order

[2] See Exercise 3.16 (5) and (10) for verifications that these are order relations.

relation, we may use other symbols that suggest directionality, such as $\leq, \preceq, \sqsubseteq, \trianglelefteq$, etc. Naturally, we write

$$a < b \iff a \leq b \text{ and } a \neq b, \tag{3.6}$$

for the *strict order* associated with an order relation $\leq$, with similar symbols for other orders, such as $\prec, \sqsubset, \triangleleft$, etc., if needed. Note that the strict order induced by a partial order is not reflexive, but it is transitive and antisymmetric.

Example 3.44 (Lexicographic Order). Let X be the set of words over the usual 26 letters roman alphabet A, that is, $W \in X$ is a finite string of symbols of A. Note that A is equipped with a total order (namely, $a \leq b \leq c \leq \ldots \leq z$). We define on X the *lexicographic order* $\preceq$ induced by $\leq$: First to compare two words $W_1 = (x_1x_2\ldots x_k)$ and $W_2 = (y_1y_2\ldots y_k)$ of the same length, let $W_1 \preceq W_2$ if $x_i \leq y_i$ for the first $i \in \{1,\ldots,k\}$ where x_i and y_i differ. If now W_1 and W_2 do not have the same length, the shortest word is completed by "blank" spaces (or any designated symbol) that is considered smaller than every element of A. Hence

$$\text{duress} \preceq \text{during}$$

because $e \leq i$ in A, and

$$\text{dye} \preceq \text{dyed}$$

because the blank space added at the end of "dye" to obtain a string of length 4 is considered less than any letter. The relation $\preceq$ is a total order on X (see Exercise 3.46 below).

In the previous example, the order $\preceq$ on words is induced by the order $\leq$ on the alphabet. This construction can be generalized:

Example 3.45 (Generalized Lexicographic Order). Let $(A,\leq)$ be a finite poset, that we think of as the *alphabet*. Let X denote the set of finite sequences (including the empty sequence) of elements of A, that we think of as the *set of words* formed using the alphabet A. The set X of words is naturally endowed with a binary operation $\frown$ of *concatenation* of words. For instance, if $W = abc$ and $G = fg$, then

$$W \frown G = abcfg.$$

We say that a word P is a *prefix* of another word W if there is a word S with $P \frown S = W$. Define now the relation $\preceq$ on X by $W \preceq Z$ if either W is a prefix of Z or there exists (possibly empty) words U, V, and R and letters $a < b$ in A with

$$W = U \frown a \frown V \text{ and } Z = U \frown b \frown R.$$

Exercise 3.46. Show that the relation $\preceq$ of Example 3.45 is an order relation. Show that if the order $\leq$ on A is total, so is $\preceq$.

Example 3.47 (Product and Lexicographic Order). If $(X,\leq_X)$ and $(Y,\leq_Y)$ are posets, then $X \times Y$ can be ordered by the *product order* $\preceq$ defined by

$$(x_1,y_1) \preceq (x_2,y_2) \iff x_1 \leq_X x_2 \text{ and } y_1 \leq_Y y_2.$$

It is easily verified that $\preceq$ is an order relation on $X \times Y$ (see Exercise 3.49 below). Note that even if $\leq_X$ and $\leq_Y$ are total orders, the product order $\preceq$ is not. Take, for instance, $X = Y = \mathbb{R}$ equipped with its usual order $\leq$. In the product order $(1,2)$ and $(2,1)$ are not comparable.

Alternatively, one may order $X \times Y$ with the *lexicographic product order* $\preceq_\ell$ defined by

$$(x_1,y_1) \preceq_\ell (x_2,y_2) \iff x_1 <_X x_2 \text{ or } (x_1 = x_2 \text{ and } y_1 \leq_Y y_2).$$

Note that

$$(x_1,y_1) \preceq (x_2,y_2) \Longrightarrow (x_1,y_1) \preceq_\ell (x_2,y_2),$$

but unlike the product order $\preceq$, the lexicographic product order $\preceq_\ell$ is a total order if $\leq_X$ and $\leq_Y$ are both total orders.

Remark 3.48. The usual ordering of numbers as in Example 3.15 or Exercise 3.16 (4), or the lexicographic order of the dictionary, as in Example 3.44, are familiar total order relations. But our minds more generally process information often by ordering it, assigning value to various experiences in order to compare them, in order to be able to decide that "this is better than that."

This is complicated when considering elements that are not comparable for the "natural" order relation. For instance, when faced with a choice, we evaluate the pros and cons of the alternatives, hence comparing them from different viewpoints, from the perspective of different order relations. Imagine, for example, that you are shopping for a new car. Car 1 is better than car 2 when it comes to safety, but car 2 is better than car 1 when it comes to mileage. In considering two criteria (and of course, you typically consider more than two in practice), we try to compare pairs where the first coordinate is about safety and the second about mileage. Hence we compare in a product of two sets. Yet, in the product order, many elements are not comparable. Our cars 1 and 2 are not. The product order yields a decision only in the case of an obvious choice, that is, when one car is superior to the other with respect to all your criteria.

In using the lexicographic product order instead, we implicitly place a total order on the set of criteria: the first coordinate is the deciding criterion: if a car is better than the other in this category, this determines your choice. If they are equal in this category, the second criterion becomes the deciding factor, and so on. This can be meaningful in some cases, but often, there isn't a single determining criterion that would be enough to offset all differences in other categories. Say, if price is your primary criterion, it cannot be in practice the only determining factor if the cheapest car you are considering is junk in every other category. . .

Hence it is a standard feature of our experience that many elements that we are trying to order are not comparable in the "natural" order; this is what creates dilemmas. Thus as a (of course drastically simplified) model for this mental ordering process, order relations typically have non-comparable elements.

Exercise 3.49. Show that the product order $\preceq$ and lexicographic product order $\preceq_\ell$ defined in Example 3.47 are indeed order relations on $X \times Y$.

Definition 3.50. Let $(X, \leq)$ be a poset, and let $a, b \in X$. We say that a is an *immediate predecessor of* b or that b is an *immediate successor of* a if $a < b$ and there is no element $c \in X$ with $a < c < b$.

For finite posets, the order structure is determined by immediate successors or predecessors and can thus more easily be depicted:

Definition 3.51 (Hasse Diagrams). The *Hasse diagram* of a finite poset $(X, \leq)$ is a graph without loop in which elements of X are vertices and an edge directed upward from x to y is drawn if y is an immediate successor of x.

The convention in a Hasse diagram is that the order increases as we move up in the picture. For instance,

Example 3.52. The Hasse diagram of the poset $X = \{1,2,3,4,6,8,9,12,18,24\}$ ordered by

$$x \leq y \iff x|y$$

is

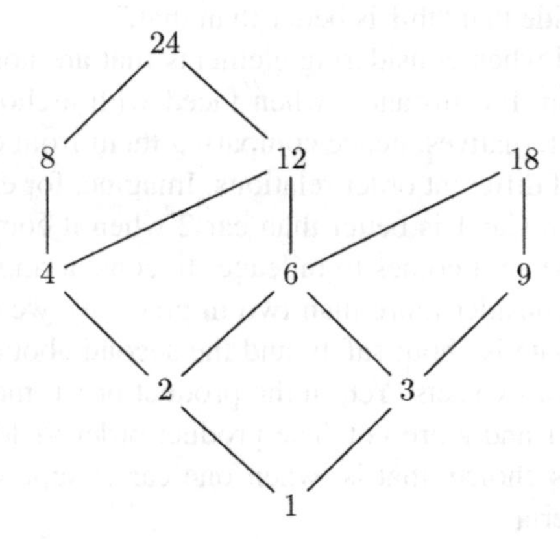

because 2 and 3 are the immediate successors of 1, 4, and 6 are the immediate successors of 2, 6, and 9 of 3, and so on.

Remark 3.53. Note that in giving a Hasse diagram, we implicitly assume that it represents an order relation. This is why loops and arrows coming from transitivity can be omitted, in order to make the graph more readable. Indeed, the graph of the relation in the sense of Remark 3.8 easily becomes too crowded to be useful. Consider, for instance, the full relational graph corresponding to the Hasse diagram above, which contains so much information that it is nearly useless:

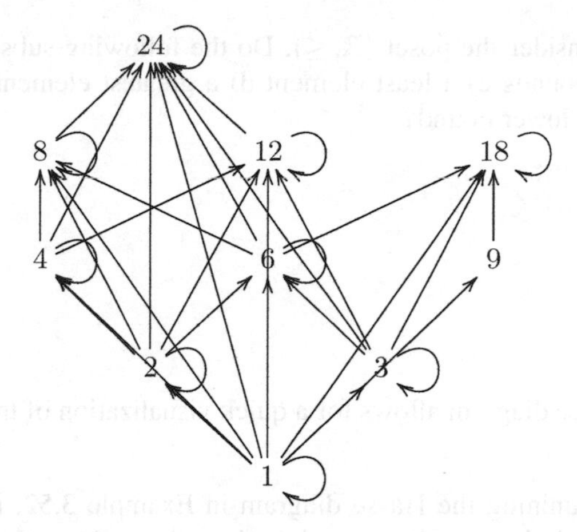

Exercise 3.54. If this is the Hasse diagram for an order relation R, write out explicitly the relation:

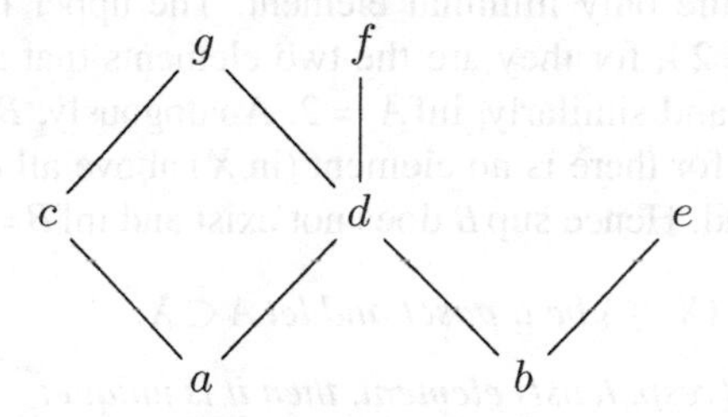

Definition 3.55. If $(X, \leq)$ is a partially ordered set, and $A \subset X$ then $w \in X$ is an *upper bound of* A if $w \geq a$ for all $a \in A$ and a *lower bound of* A if $w \leq a$ for all $a \in A$.

The *greatest element* of $A \subset X$ is an element of A that is also an upper bound of A. The *least element of* $A \subset X$ is an element of A that is also a lower bound of A.

The *greatest lower bound* or *infimum* of $A \subset X$ is the greatest element of the set of all lower bounds of A, and is denoted $\inf A$ or $\inf_{a \in A} a$ or $\bigwedge A$ or $\bigwedge_{a \in A} a$. The *least upper bound* or *supremum* of $A \subset X$ is the least element of the set of all upper bounds of A, and is denoted $\sup A$ or $\sup_{a \in A} a$ or $\bigvee A$ or $\bigvee_{a \in A} a$.

An element $m \in X$ is *maximal* if

$$m \leq x \Longrightarrow m = x$$

and *minimal* if

$$x \leq m \Longrightarrow m = x.$$

For a given subset of a given poset, and for each of the notions listed in Definition 3.55, there may or may not exist elements satisfying the property. A lower or upper bound for a set A (when it exists) may or may not belong to the set A. For instance, in $(\mathbb{R}, \leq)$, if $A = [0, 1)$, then 5 is an upper bound that does not belong to A. The least upper bound is 1, which does not belong to A either. A does not have a greatest element, but has a least element 0, which is also the greatest lower bound.

Exercise 3.56. Consider the poset $(\mathbb{R},\leq)$. Do the following subsets have a) lower bounds b) upper bounds c) a least element d) a greatest element e) a least upper bound f) a greatest lower bound?

1. $\mathbb{N}$
2. $\mathbb{Z}$
3. $(1,2)$
4. $[1,2)$
5. $(1,2]$
6. $[1,2]$.

Note that a Hasse diagram allows for a quick visualization of the order structure. For instance,

Example 3.57. Examining the Hasse diagram in Example 3.52, it is clear that 24 and 18 are maximal elements because there is nothing above them, but that there is no greatest element (no element above every other). On the other hand, 1 is the least element, hence the only minimal element. The upper bounds of the subset $A=\{4,6\}$ are 12 and 24, for they are the two elements that are above both 4 and 6. Hence $\sup A=12$, and similarly, $\inf A=2$. Analogously, $B=\{4,6,9\}$ does not have an upper bound, for there is no element (in X) above all three numbers, and 1 is the only lower bound. Hence $\sup B$ does not exist and $\inf B=1$.

Proposition 3.58. *Let $(X,\leq)$ be a poset and let $A\subset X$.*

1. *If A has a greatest (resp. least) element, then it is unique;*
2. *If A has a greatest lower bound (resp. least upper bound), then it is unique;*
3. *If A has a greatest element (resp. least element) it is also the least upper bound (resp. greatest lower bound) and the unique maximal (resp. minimal) element;*
4. *If A has a least upper bound (resp. greatest lower bound) m and $m\in A$, then m is the greatest (resp. least) element of A.*

Proof. (1): If m and b are two greatest elements of $A\subset(X,\leq)$, then $m\geq b$ because $b\in A$ and m is a greatest element, and similarly, $b\geq m$ because $m\in A$ and b is a greatest element of A. But $\geq$ is antisymmetric, so that $m=b$.

(2): Since the greatest lower bound is the greatest element of the set of lower bounds, and, by (1), greatest elements are unique whenever they exist, we conclude that greatest lower bounds are unique whenever they exist.

(3): If m is the greatest element of A, then m is an upper bound for A and $m\in A$, so that any other upper bound u for A satisfies $u\geq m$. Thus m is the least upper bound.

(4): If $m=\sup A\in A$, then $m\geq a$ for all $a\in A$ because m is an upper bound for A. Thus m is the greatest element of A. □

However, minimal and maximal elements are not unique in general.

Exercise 3.59. Consider the poset $(\mathbb{N},|)$. Let $A=\{1,2,3,6\}$ and $B=\{4,6\}$.

1. Give, if they exist, the greatest and least element of A and B. Justify your answers.

2. Give, if they exist, the maximal and minimal elements of A and B. Justify your answers.
3. What are, if any exist, the lower bounds and upper bounds of A and of B? Justify your answers.
4. What are, if they exist, the greatest lower bound and least upper bound of A and of B?

Remark 3.60. Of course, you should recognize that the least upper bound of a set $C \subset \mathbb{N}$ for the division order $|$ is the least common multiple of elements of C, while the greatest lower bound of C is the greatest common divisor of elements of C.

Note that as a result, every finite subset of $\mathbb{N}$ has a least upper bound (the least common multiple) and a greatest lower bound (the greatest common divisor). In other words, the poset $(\mathbb{N}, |)$ is a *lattice*, that is, a poset in which every non-empty finite subset has a greatest lower bound and a least upper bound. On the other hand, the (infinite) subset of prime numbers does not have any upper bound (there is no common multiple of all prime numbers), hence no least upper bound. In other words, $(\mathbb{N}, |)$ is not a complete lattice (see Definition 3.63 below).

Exercise 3.61. Let $X = \{a,b,c,d\}$ and consider the poset $(\mathbb{P}X, \subset)$. Let $A \subset \mathbb{P}X$ be

$$A = \{\{a,b\},\{a,c\},\{a,b,c\}\}.$$

1. What are the maximal and minimal elements of A?
2. Does A have a greatest element? a least element?
3. What are the upper bounds of A? What are the lower bounds of A?
4. What are the greatest lower bound and least upper bound of A?

Exercise 3.62. Let X be a set. Show that every subset of the partially ordered set $(\mathbb{P}X, \subset)$ has a least upper bound and a greatest lower bound. Describe them.

In other words, $(\mathbb{P}X, \subset)$ is a complete lattice in the following sense:

Definition 3.63. A poset X is a *complete lattice* if every subset of X has a least upper bound.

Exercise 3.64. Show that if X is a complete lattice then

1. X has a greatest element;
2. every subset of X has a greatest lower bound;
3. X has a least element.

Remark 3.65. Note that $\mathbb{R}$ is not a complete lattice, for it does not have a least or a greatest element. However, $[0,1]$ is a complete lattice: the fact that bounded subsets of the reals have a least upper bound is a key property of the real numbers that we will revisit when we examine a construction of the real line.

Example 3.66 (Pointwise Order). Consider the set F of functions from a closed interval $[a,b]$ and valued in $\mathbb{R}$. Define on this set the *pointwise order* defined by

$$f \leq g \iff \forall x \in [a,b]\ f(x) \leq g(x) \text{ (in } \mathbb{R}). \tag{3.7}$$

Exercise 3.67. For the pointwise order of Example 3.66,

1. Show that (3.7) defines an order relation on F;
2. Is the pointwise order a total order on F? Justify.
3. Show that the subset C of F composed of functions from $[a,b]$ to $[0,1]$ is a complete lattice.
4. What is the greatest and least element of C?

Exercise 3.68. Let $(\mathbb{Q},\leq)$ be the set of rational numbers endowed with the usual order relation. Does the set $A=\{1-\frac{1}{n}: n\in\mathbb{N}\}$ have a greatest element? A least upper bound? If they exist, what are they. Justify your answers. Consider the same questions for $B=\{r\in\mathbb{Q}: r\leq 0\vee r^2<2\}$.

We have seen in Exercise 3.64 that a complete lattice has a greatest element, denoted here $\top$ (for "top"), and a least element, denoted here $\bot$ (for "bottom").

A function $f: X\to Y$ between posets $(X,\leq)$ and $(Y,\sqsubseteq)$ is *increasing* or *order-preserving* if

$$x_1\leq x_2 \Longrightarrow f(x_1)\sqsubseteq f(x_2)$$

for every $x_1,x_2\in X$. It turns out that order-preserving self-maps of complete lattices have fixed points—a fact that we will use in the next chapter to prove the Cantor-Bernstein Theorem. This follows from the following stronger result:

Theorem 3.69 (Knaster-Tarski Fixed Point Theorem). *If $(L,\leq)$ is a complete lattice and $f: L\to L$ is order-preserving, then the set*

$$F=\{x\in L: f(x)=x\}$$

of fixed points of f is a complete lattice, and thus is not empty.

We only show here that $F\neq\emptyset$, which is all we need for our application to Theorem 4.5. We prove the full statement in Appendix A.2.

Proof. Consider the set

$$D=\{x\in L: x\leq f(x)\}.$$

Since L is a complete lattice its supremum $s=\bigvee D$ exists. Of course, $x\leq s$ for all $x\in D$, and f is order-preserving so

$$\forall x\in D\,(x\leq f(x)\leq f(s))$$

so that $f(s)$ is an upper bound for D. Thus $s\leq f(s)$ because s is the least upper bound. Hence $s\in D$. On the other hand,

$$f[D]\subset D.$$

To see this, note that if $x\in D$, that is, $x\leq f(x)$, then $f(x)\leq f(f(x))$ because f is order-preserving, so that $f(x)\in D$. In particular, $f(s)\in D$ so that $f(s)\leq\bigvee D=s$ and $s\leq f(s)$ because $s\in D$. Thus $s=f(s)\in F$. □

To conclude this section, let me mention that the *principle of mathematical induction* introduced at a "naive" level in Section 2.6 has to do with a particular order property of the natural number, namely, *well-orderedness*. As a result, this principle can be generalized. See Section A.3 for details.

Additional Exercises (Homework)

Exercise 3.70. Is the relation R on $\mathbb{N}$ defined by nRp if there is an integer $k \geq 0$ with $p = 2^k n$ an order relation?

Exercise 3.71. Recall that the modulus $|z|$ of a complex number $z = a + ib$ is $|z| = \sqrt{a^2 + b^2}$. Is the relation R on $\mathbb{C}$ defined by

$$z_1 R z_2 \iff |z_1| \leq |z_2|$$

an order relation?

Exercise 3.72. Let R be a relation on $X = \{a, b, c, d\}$ defined by

$$R = \{(a,a), (b,b), (c,c), (d,d), (a,b), (a,c), (b,d), (a,d)\}.$$

1. Verify that R is an order relation.
2. Draw its Hasse diagram.
3. Does X have a least and a greatest element for this order? What are the minimal and maximal element.
4. Let $A = \{b, c\} \subset X$. Give, if they exist, $\inf A$ and $\sup A$ for this order.

Exercise 3.73. Let $X = \{1, 2, 3, 4, 5, 6\}$ be ordered by the order relation with Hasse diagram

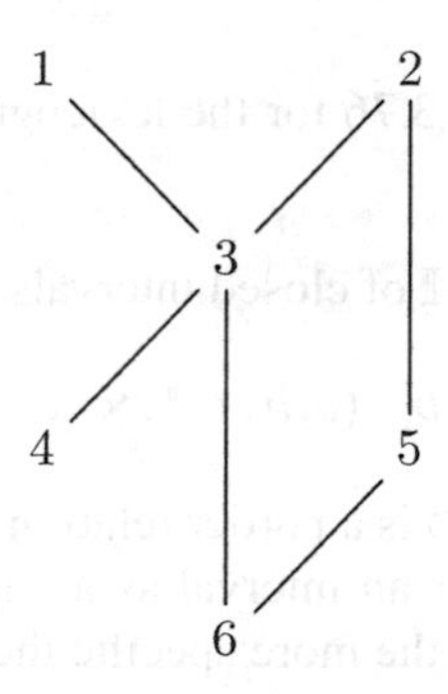

1. Find all minimal and maximal element of X;
2. Does X have a least element? A greatest element?
3. Find all totally ordered subsets of X that have at least 3 elements.

Exercise 3.74. Let $Y = \{a,b,c,d,e,f\}$ be ordered by the order relation with Hasse diagram

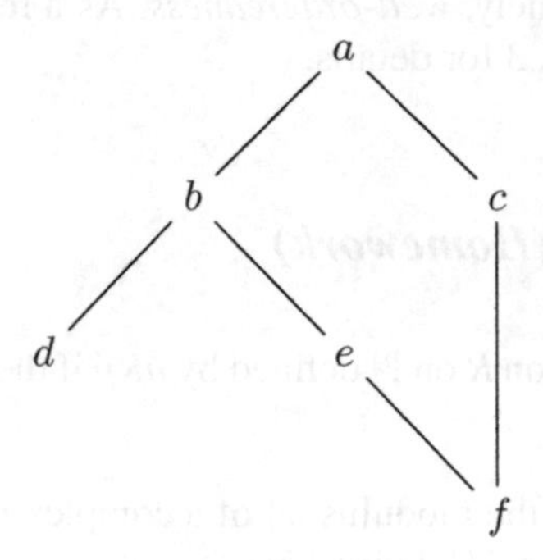

1. Find all minimal and maximal elements;
2. Does Y have a least element? a greatest element?
3. Let $A = \{d,e,c\} \subset Y$. Find $\sup A$ and $\inf A$ if they exist.

Exercise 3.75. Show that in a powerset $\mathbb{P}X$ ordered by inclusion $\subset$, where X is a non-empty set:

1. If $S \in \mathbb{P}X$ and $x \in X \setminus S$, then $S \cup \{x\}$ is an immediate successor of S;
2. If $x \in S \in \mathbb{P}X$, then $S \setminus \{x\}$ is an immediate predecessor of S.

Exercise 3.76. Let $(X, \leq_X)$ and $(Y, \leq_Y)$ be the posets of Exercises 3.73 and 3.74. Consider the product order relation as defined in Example 3.47. Compare the following elements of $X \times Y$ or say if they are not comparable:

1. $(4,b)$ and $(2,e)$;
2. $(3,a)$ and $(6,f)$;
3. $(5,d)$ and $(1,a)$;
4. $(6,e)$ and $(2,b)$.

Exercise 3.77. Repeat Exercise 3.76 for the lexicographic product order on $X \times Y$, as defined in Example 3.47.

Exercise 3.78. Consider the set $\mathbb{I}$ of closed intervals in $\mathbb{R}$, that is,

$$\mathbb{I} = \{[a,b] : (a,b) \in \mathbb{R} \times \mathbb{R},\ a \leq b\}.$$

1. Show that reverse inclusion $\supset$ is an order relation on $\mathbb{I}$. We call it the *information order* because if we consider an interval as an approximation of a number ([3]), then the smaller the interval, the more specific the information on the number.
2. Does $\mathbb{I}$ have minimal elements? maximal elements?

---

[3] With the understanding that the actual number we are after is in that interval, and that the interval represents the partial information we have on that number: it is between a and b.

3. Let

$$\mathscr{A} = \left\{[-1,1], \left[-\frac{1}{2},\frac{1}{2}\right], [0,2], \left[0,\frac{2}{3}\right], \left[\frac{1}{4},\frac{2}{3}\right]\right\}.$$

Describe

a. the maximal and minimal elements of $\mathscr{A}$ in $(\mathbb{I},\supset)$;
b. the least upper bound and greatest lower bound of $\mathscr{A}$ in $(\mathbb{I},\supset)$;
c. Does $\mathscr{A}$ have a greatest or a least element. Explain.

4. Let $(X,\leq)$ be the set $\mathscr{A}$ above with the order induced by that of $\mathbb{I}$.

a. Draw the Hasse diagram.
b. Consider the following subsets of X:

$$A = \{[-1,1],[0,2]\},\ B = \left\{[-1,1], \left[-\frac{1}{2},\frac{1}{2}\right], \left[0,\frac{2}{3}\right], \left[\frac{1}{4},\frac{2}{3}\right]\right\}.$$

i. Describe, if any, the upper and lower bounds of A and of B (in X);
ii. Describe, if any, the least upper and greatest lower bound of A and of B (in X);
iii. Describe, if any, the least and greatest elements of A and of B;
iv. Describe, if any, the minimal and maximal elements of A and of B.

3.3 Equivalence

Recall that an *equivalence relation* on a set X is a relation on X that is reflexive, symmetric, and transitive.

Exercise 3.79. If X is the set of US nationals and $R \subset X \times X$ is defined by xRy if x and y have the same age (rounding to the nearest year), what kind of relation is R?

Exercise 3.80. Let X be the set of all lines in the plane, and for $x,y \in X$ let xRy if x and y are parallel and xSy if x and y are perpendicular. Are R and S equivalence relations?

Definition 3.81. If R is an equivalence relation on X, then for each $x \in X$, the set

$$R(x) = \{y \in X : xRy\}$$

is called *equivalence class of x*.

Lemma 3.82. *Distinct equivalence classes are disjoint.*

Proof. Let R be an equivalence relation on X. If $R(t) \neq R(x)$, then there is an element of one equivalence class that is not in the other, say, there is $z \in R(t) \setminus R(x)$. Then $R(t) \cap R(x) = \emptyset$. Assume to the contrary that there is $y \in R(t) \cap R(x)$. By symmetry of R, we have yRt and tRz, so that yRz by transitivity. But xRy, so that xRz by transitivity, in contradiction to $z \notin R(x)$. □

Note that moreover, each $x \in X$ belongs to at least one equivalence class for $x \in R(x)$ by reflexivity. Hence

$$X = \bigcup_{x \in X} R(x)$$

and any two equivalences classes are either equal or disjoint by Lemma 3.82. Thus an equivalence relation determines a *partition of X* into equivalence classes.

Definition 3.83. A set $\mathscr{P} \subset \mathbb{P}X$ of subsets of X with $\bigcup_{P \in \mathscr{P}} P = X$ and for every $P_1, P_2 \in \mathscr{P}$,

$$P_1 \neq P_2 \Longrightarrow P_1 \cap P_2 = \emptyset,$$

is called a *partition* of X.

Exercise 3.84. Show that, conversely, if $\mathscr{P}$ is a partition of X then

$$x \sim_{\mathscr{P}} y \iff \exists P \in \mathscr{P} \ (\{x, y\} \subset P)$$

is an equivalence relation, the equivalence classes of which are the elements of $\mathscr{P}$.

Definition 3.85. If R is an equivalence relation on X, the set

$$X/R = \{R(x) : x \in X\}$$

of equivalence classes for R is called *quotient of X by R.* The map $q_R : X \to X/R$ that associates with each point of x its equivalence class $q_R(x) = R(x)$ is onto and is often called *canonical surjection* associated with R.

Thus an equivalence relation (equivalently, a partition) determines a surjection. Conversely, every surjective map determines an equivalence relation on its domain for which it is (essentially, in a sense to be explained below) the canonical surjection:

Theorem 3.86. *Let $f : X \to Y$ be an onto map. Then the relation on X defined by*

$$x \sim_f t \iff f(x) = f(t)$$

is an equivalence relation, and the map $\widehat{f} : X/\sim_f \to Y$ defined by $\widehat{f}(q(x)) = f(x)$ is a well-defined bijection.

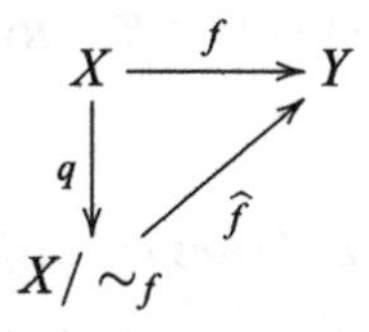

Proof. The relation $\sim_f$ is clearly reflexive, symmetric, and transitive, hence an equivalence relation. To see that $\widehat{f}$ is well-defined, note that if $x, t \in X$ with $q(x) = q(t)$ then $x \sim_f t$, that is, $f(x) = f(t)$. Hence $\widehat{f}(q(x)) = \widehat{f}(q(t))$. To see that $\widehat{f}$ is a bijection, note that for every $y \in Y$, there is $x \in X$ with $f(x) = y$ because f is onto, so that $\widehat{f}(q(x)) = y$ and $\widehat{f}$ is onto. Moreover, if $\widehat{f}(q(t)) = \widehat{f}(q(x))$, then $f(t) = f(x)$, that is, $x \sim_f t$, and thus $q(t) = q(x)$. Thus $\widehat{f}$ is one-to-one. □

Remark 3.87. Let us try to interpret this theorem: an onto map defines an equivalence relation for which the set $\{f^{-1}[y] : y \in Y\}$ of its fibers is the corresponding partition of X into equivalence classes. Moreover, there is a bijection between the quotient set and Y. In other words, to each equivalence class corresponds exactly one element of Y and each element of Y corresponds to exactly one equivalence class. One way to think of this is that each equivalence class can uniquely be "given a name" in Y, so that we can essentially identify $X/\sim_f$ and Y. To understand this identification, consider, for instance, the equivalence relation of Exercise 3.79.

The equivalence class of a given US person is the set of all US nationals who have the same age as that person. The quotient set is the set of such equivalence classes. Hence, in the quotient set, the whole equivalence class is represented by a single point (note that therefore, in passing to the quotient, we "forget" about all differences between people, but their age).

Theorem 3.86 helps us think of this point (an equivalence class in the set of equivalence classes—not an intuitive thing to think about) in a more meaningful way: Let X be the set of US nationals, Y the set of all ages of US nationals (a finite subset of that of integers), and let f be the map that associates with each US national his or her age. This map is onto and $\sim_f$ is the equivalence relation at hand in Exercise 3.79. The map $\widehat{f}$ attaches a meaningful label to each point of the quotient set: formally, an element of the quotient is an equivalence class, "representing" all of its members, who all have the same age, say n. The map $\widehat{f}$ simply "names" this element of the quotient set n. In practice, we think of the quotient set as Y, where the common age n of an age group is what represents the group, rather than a formal equivalence class in a set of equivalence classes. Via this identification of $X/\sim_f$ with Y, we interpret f as the canonical surjection for $\sim_f$. In this example f associates with each person in X his or her age in the (interpreted as opposed to formal) quotient set.

Remark 3.88. One more comment is in order before revisiting previous examples of equivalence relations from this perspective. We have seen that, in some contexts, it makes sense to identify two sets that admit a bijection between them. For instance, $\{\bigcirc, \square, \triangle\}$ and $\{1,2,3\}$ are two different sets because they have different elements, but the difference is only the names we give to each elements, and how we want to interpret what these labels represent. From one set to another, there is only a change of labels. In many contexts, we can think of these two sets as two sides of the same coin, as the same set, viewed from two different perspectives.

In fact, you are not foreign to the notion of identifying objects that are not exactly the same formally. For instance, when you look at two congruent triangles in the plane, you consider them to be the same, because one is the image of the other under a rigid transformation. Hence you look at objects up to rigid transformations. In the context of Euclidean geometry, *isomorphisms* are rigid transformations, and we look at the objects of study up to isomorphisms. Two *isomorphic* objects (that is, two objects that are image of one another under an isomorphism) in this context represent the same thing in two different positions in the plane.

As mentioned in Remark 1.84, advanced mathematics is concerned in large part with the study of sets with additional structures, such as a group structure, a vector space structure, a topological structure, etc. ([4]). Transformations that "respect" the structure in a certain sense (group homomorphisms for groups, linear maps for vector spaces, continuous maps for topological spaces) are of particular interest and are often called *morphisms*. Isomorphisms are usually bijective maps (hence maps with an inverse as in Definition 1.110) that are morphisms in both directions. Hence they bijectively transport the structure from one object to the other, so that two isomorphic objects are essentially the same *from the point of view of the studied structure*. You may think of a collection of isomorphic objects as the same object viewed from different standpoints, thus helping you to understand its true nature.

In the context of sets, the "structure" is very simple: it is the list of elements. Isomorphisms between two sets are the maps that transport one list onto the other in a way that can be reverse: they are the bijective maps. Hence, we study sets to a large extent up to bijections. We will see in Chapter 4 that the "change of perspective on the same object" provided by a bijection can be surprising at first.

At any rate, you should re-read Theorem 3.86 in light of these remarks. As you progress through mathematics courses, you will see several theorems akin to Theorem 3.86, stating what morphisms f can be interpreted as the canonical surjection onto a quotient set equipped with the appropriate structure.

Let us now revisit examples of equivalence relations we have seen so far, from these different perspectives (equivalence relation, partition, onto map).

Example 3.89. Consider

1. the equivalence relation R of Exercise 3.80. The equivalence class of a given line L in the plane is the set of all lines parallel to L. In the quotient set, the whole equivalence class is identified with a point, that "represents" all of these parallel lines. Hence we can think of this point of the quotient as the common direction of these lines, and of the quotient set as the set of possible directions. In passing to the quotient set, we "forget" about all differences between lines but their directions. The quotient set is the set of possible directions. It can be identified with a half circle centered at the origin of the plane with only one endpoint of the diameter ([5]), each point representing a direction of a line through the origin and this point:

---

[4]As we are making a very general meta-mathematical comment, it is of little importance if you know the definitions of these notions. You will get acquainted to them in time, but I hope the larger point to be helpful nevertheless.

[5]Note that in introducing this interpretation, we implicitly consider a map f that associates with each line in the plane the end point of a unit directional vector, and recast the situation via Theorem 3.86. Of course, we could give a different yet equivalent interpretation, via another map, for instance, a map g onto $\mathbb{R}\cup\{\infty\}$ associating with each non-vertical line its slope and ∞ with vertical lines. Hence there is more than one way to interpret a quotient set via Theorem 3.86.

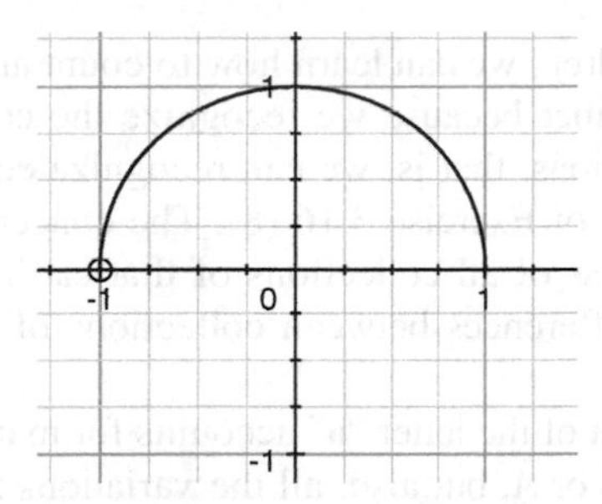

2. the equivalence relation of Example 3.14 considered on $\mathbb{N}$ (two natural numbers are equivalent if they are either both even or both odd). The partition of $\mathbb{N}$ into equivalence classes has only two elements: the set of even natural numbers and the set of odd natural numbers. Hence the quotient set has only two points that we can think of $\{0,1\}$, the possible remainders in the division by 2. Each even number is identified with 0 (the remainder in its division by 2) in the quotient set, and each odd number is identified with 1 (the remainder in its division by 2) in the quotient set. Hence, in passing to the quotient, we "forget" all differences between numbers but their parity (or equivalently, their remainder in the division by 2).
3. the equivalence relation of Exercise 3.16 (8). The equivalence class of a given finite set $A \in X$ is the set of all sets in X that have the same cardinality as A. In the quotient set, all these sets are identified to a single point that thus represents this common cardinality.

Here is an example of fundamental importance in arithmetic.

Exercise 3.90. Let p be an integer greater than 1. Define on $\mathbb{Z}$ the relation

$$x \equiv_p y \iff p \text{ divides } x - y.$$

Show that $\equiv_p$ is an equivalence relation. If $x \equiv_p y$, we say that x and y are *congruent modulo p*.

Describe the equivalence classes, the quotient set, and the canonical surjection.

Remark 3.91. An important point to note is that while passing to the quotient set for an equivalence relation might seem like an abstract construction, it is nothing but a formal model for a very basic thought process that we use all the time without giving it a thought.

Equivalence relations are a model for identifying patterns, and passing to the quotient set is a model for identification of alike patterns in order to interpret the world in simplified terms. One may argue that these processes are fundamental mental feature of human beings.

For instance, as children, we can learn how to count and make sense of the concept of an abstract number because we recognize the common trait "cardinality" between various collections, that is, we can recognize equivalence in the sense of the equivalence relation of Exercise 3.16 (8). The concept "3" is abstract: it is really the equivalence class of all collections of that cardinality. We make sense of "3" by forgetting all differences between collections of various objects, but their cardinality.

Similarly, the concept of the letter "a" accounts for many calligraphic variations: to start with a or *a* or A or *A*, but also, all the variations in handwriting. Our brain "forgets" all these differences and identifies all these symbols by considering them *equivalent*. We can communicate through handwritten symbols only because we identify (via a relation that is not as tightly defined as those we consider in the more controlled and limited environment of mathematics, but an equivalence relation is a good model for it) many variations of a symbol to a single *ideal conceptual* version of the symbol: the one in the quotient set that *represents* all the variants.

More generally, *concepts,* even the most mundane, are almost always standing for a spectrum of interpretations and/or experiences. In other words, the concept *is* the equivalence class for a collection of related but different individual meanings and/or experiences: it *is* what is common in that collection from the perspective of an untold equivalence relation. Take the concept "table," for instance. There are many instances of tables of various kinds, and even if we fix a specific table, each individual experience of that table is different ([6]). Yet there is a single concept "table" that your mind associates with each instance of table, or experience thereof, that is distinct from individual instances, yet encompasses them all, that is, the concept "represents" a whole spectrum of possible incarnations, like a point of a quotient set "represents" a range of possibilities: all the elements of the equivalence class it represents.

This observation is, in my opinion, more than a vague analogy. Our brain is wired to identify patterns (equivalence relations!) and then "think them" at a simplified conceptual level, that of the quotient set. Of course, in that context, the ambient set and the exact nature of the equivalence relation are not as clearly and formally definable as in the context of mathematics. But I hope to have convinced you that equivalence relations and quotient sets are but a mathematical formalization (hence an idealized model for) a very common thought process. Hence it should not be surprising that *passing to the quotient* for an equivalence relation is one of the most fundamental and ubiquitous constructions in mathematics.

Remark 3.92. So far, we have seen several ways to think of an equivalence relation R on X: in terms of the relation itself, which typically can be thought of as xRt if and only if x and t share a particular property. Equivalently, in terms of a partition of a

[6] For an introductory discussion of the difficulties of identifying an object with the experience one may have of it, read, for instance, [26, Chapter I].

set, where xRt if and only if x and t are in the same element of the partition—in other words, elements of X are divided in groups (elements of the partition) according to the property examined in the previous viewpoint. Finally, we noted that the partition can always be thought of as the partition into fibers

$$\{f^{-1}[y] : y \in Y\}$$

of a surjective map $f : X \to Y$, where Y can be thought of as the quotient set. Here xRt if $f(x) = f(t)$, that is, f associates with an element of X the property examined in the first viewpoint.

Moreover, we can also give a geometric interpretation to the quotient set. Indeed, elements of the same equivalence class are identified in the quotient set. Geometrically, that corresponds to gluing together elements that are equivalent to each other.

Example 3.93. Consider the map $f : [0,1] \to S_1$ defined by $f(t) = e^{2\pi it}$, where $S_1 = \{z \in \mathbb{C} : |z| = 1\}$ is the unit circle in the complex plane. Note that f is onto, that the restriction of f to $(0,1)$ is a bijection onto $S_1 \setminus \{1\}$ and that $f(0) = f(1) = 1$. Geometrically, this quotient map identifies, hence glues together, the two ends of the interval $[0,1]$ to form the circle. Hence the circle is a geometric realization of the quotient set $[0,1]/\sim_f$.

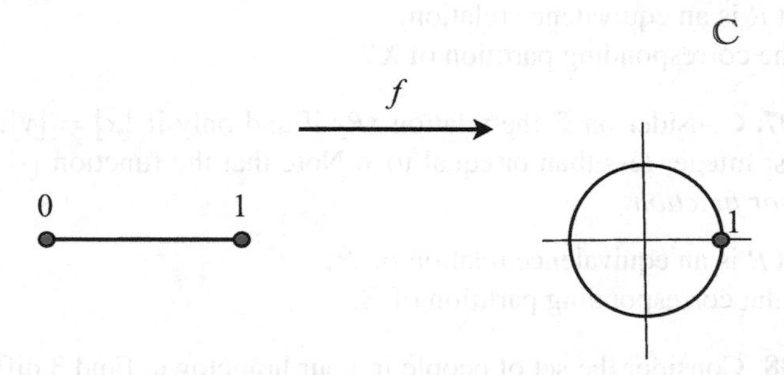

Example 3.94. Consider the unit square $[0,1] \times [0,1]$. Consider the equivalence relation $\sim$ of $[0,1] \times [0,1]$ that coincides with the identity in the interior of the square and

$$(0,y) \sim (1,y) \text{ for each } y \in [0,1], \tag{3.8}$$

$$(x,0) \sim (x,1) \text{ for each } x \in [0,1]. \tag{3.9}$$

This equivalence relation identifies opposite sides of the square (see Figure 3.1).

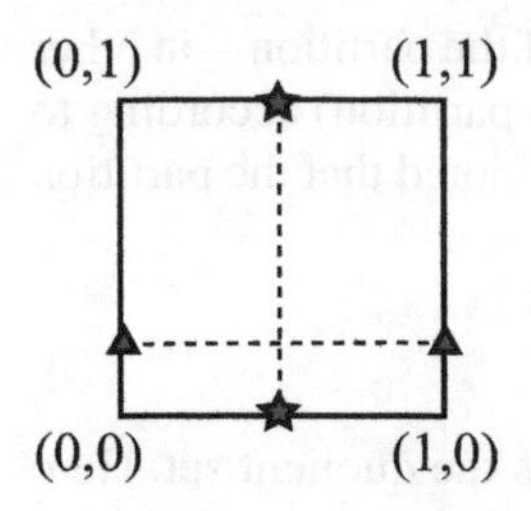

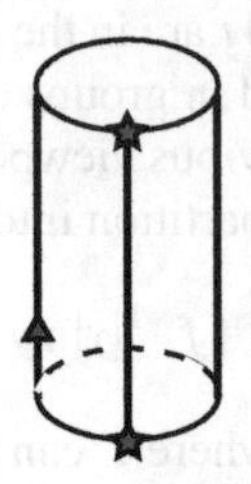
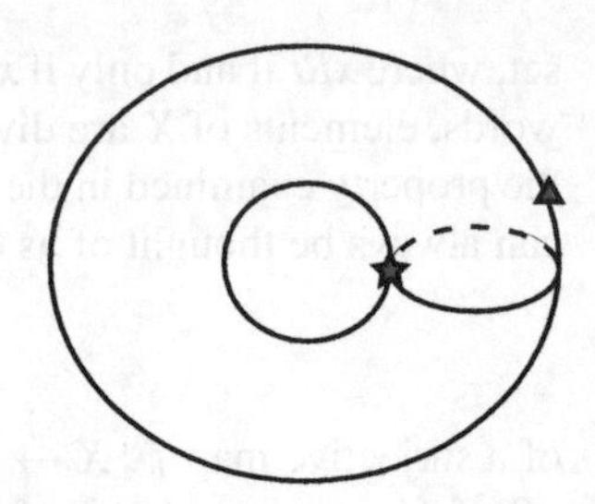

Fig. 3.1 The condition (3.8) glues together the vertical edges (the triangles in the square are identified in the cylinder), the condition (3.9) glues the horizontal edges of the square (the stars in the cylinder are identified in the torus).

Additional Exercises (Homework)

Exercise 3.95. What is the partition associated with the equivalence relation of Exercise 3.38?

Exercise 3.96. Consider $X = \{i, -i, 1, -1\} \subset \mathbb{C}$ (hence $i^2 = -1$). Let R be the relation on X defined by xRt if $xt = \pm 1$.

1. Show that R is an equivalence relation.
2. What is the corresponding partition of X?

Exercise 3.97. Consider on $\mathbb{R}$ the relation xRy if and only if $\lfloor x \rfloor = \lfloor y \rfloor$, where $\lfloor x \rfloor$ is the greatest integer less than or equal to x. Note that the function $\lfloor \cdot \rfloor : \mathbb{R} \to \mathbb{Z}$ is called the *floor function*.

1. Show that R is an equivalence relation on $\mathbb{R}$;
2. Describe the corresponding partition of $\mathbb{R}$.

Exercise 3.98. Consider the set of people in your hometown. Find 3 different ways to form a partition of that set, and interpret the partition in terms of equivalence relation. Describe the quotient set.

Exercise 3.99. Let R be a transitive and reflexive (but not necessarily antisymmetric) relation on X. Consider the relation

$$x \sim_R y \iff xRy \wedge yRx.$$

1. Show that $\sim_R$ is an equivalence relation on X and let $q : X \to X/\sim_R$ denote the canonical surjection.
2. Show that the relation $\leq_R$ on the quotient set $X/\sim_R$ defined by

$$s \leq_R t \iff \exists x \in q^{-1}[s]\ \exists y \in q^{-1}[t]\ (xRy)$$

is an order relation on $X/\sim_R$.

Exercise 3.100. Which of the following are partitions of $X = \{1,2,3,4,5,6\}$? For those $\mathscr{P}_i$ that are partitions, describe the corresponding equivalence relation, and the quotient set.

1. $\mathscr{P}_1 = \{\{1,2,3\},\{1,4,5,6\}\}$
2. $\mathscr{P}_2 = \{\{1,2\},\{3,5,6\}\}$
3. $\mathscr{P}_3 = \{\{1,3,5\},\{2,4\},\{6\}\}$

Exercise 3.101. Find all possible partitions of $X = \{a,b,c\}$. For each, describe the corresponding equivalence relation and quotient set.

Exercise 3.102. Let

$$W = \{\text{sheet,last,sky,wash,wind,sit}\}.$$

Let $\sim_l$ be the relation "has the same number of letters as" on W, and let $\sim_b$ be the relation "begins with the same letter as."

1. Check that $\sim_l$ and $\sim_b$ are equivalence relations.
2. Describe the quotient sets $X/\sim_l$ and $X/\sim_b$.

Exercise 3.103. Let R be the relation on $\mathbb{R}$ defined by xRy if $x - y \in \mathbb{Z}$.

1. Show that R is an equivalence relation;
2. Describe the equivalence classes;
3. Show that X/R is (in bijection with) the interval $[0,1)$.

Exercise 3.104. Let N be a set of integers and define on $N \times N$ the relation

$$(n,p) \sim (m,q) \iff n+q = p+m.$$

1. Show that $\sim$ is an equivalence relation.
2. If $N = \{1,2,3,4,5,6,7,8,9\}$, what is the equivalence class of $(2,5)$?

3.4 Equivalence, Order, and Sets of Numbers

Let us review basic facts on the hierarchy of sets of numbers

$$\mathbb{N} \subset \mathbb{Z} \subset \mathbb{Q} \subset \mathbb{R}.$$

Learning to count, we consider natural numbers and learn to add and multiply them, resulting in new natural numbers. In other words, we operate in the set $\mathbb{N}$ of all *natural numbers*, which is closed under addition and multiplication. We can consider the inverse operation to $+$, the subtraction, but while $5-2=3$ makes sense in $\mathbb{N}$, $3-5$ does not.

This leads to the introduction of the set $\mathbb{Z}$ of *integers*, which includes both positive and negative whole numbers. Multiplication and addition are naturally extended to $\mathbb{Z}$. Now each element of $\mathbb{Z}$ has an additive inverse (that is, $(\mathbb{Z},+)$ is a group). But many elements of $\mathbb{Z}$ have no multiplicative inverse. For instance, $2x = 1$ has no solution in $\mathbb{Z}$.

This leads to the introduction of the set $\mathbb{Q}$ of *rational numbers,* that is, quotient of integers. Formally, we define on

$$\mathbb{Z}\times(\mathbb{Z}\backslash\{0\})$$

an equivalence relation

$$(p,q)\sim(r,t) \iff pt = rq$$

and we verify that the standard operations $+$, $-$, $\times$, and $\div$ extend from $\mathbb{Z}$ to the quotient set

$$\mathbb{Q} = \mathbb{Z}\times(\mathbb{Z}\backslash\{0\})/\sim . \tag{3.10}$$

Namely, denoting $\frac{p}{q}$ the equivalence class of (p,q), we have by definition

$$\frac{p}{q} = \frac{r}{t} \iff pt = rq \tag{3.11}$$

and we can define

$$\frac{p}{q}+\frac{r}{t} = \frac{pt+rq}{tq} \text{ and } \frac{p}{q}\times\frac{r}{t} = \frac{pr}{qt}$$

on $\mathbb{Q}$. Now $(\mathbb{Q},+,\times)$ is a field, that is $(\mathbb{Q},+)$ is a group, $\times$ is commutative, has a neutral element (the number 1), and distributes over addition so that $(\mathbb{Q},+,\times)$ is a commutative ring, and every non-zero element has a multiplicative inverse. However, many equations that have "concrete" solutions do not have solutions in $\mathbb{Q}$.

For instance, the solution $\sqrt{2}$ of the equation with rational coefficients

$$x^2 = 2$$

has a "concrete" meaning, for it is the length of the diagonal of a square whose sides have length one. But $\sqrt{2} \notin \mathbb{Q}$, as seen in Proposition 2.44.

Thus we need to "complete" $\mathbb{Q}$ to include solutions of such equations. One among several ways to do this is via *Dedekind cuts* :

Definition 3.105. A *Dedekind cut*, or *cut* for short, is a non-empty proper subset of C of $\mathbb{Q}$ that is closed downward, that is,

$$c \in C \text{ and } d < c \Longrightarrow d \in C,$$

and has no greatest element, that is,

$$c \in C \Longrightarrow \exists d \in C\ (d > c).$$

It turns out that $\mathbb{R}$ can be constructed as the set of Dedekind cuts. Roughly speaking, we are going to represent a real number r by the set of all rationals that are smaller than r. As one can easily see, the set C_r of rationals smaller than a given real number r is a cut, and every cut C has a unique least upper bound r in $\mathbb{R}$ and it satisfies $C = C_r$. For instance, $\sqrt{2}$ is represented by the cut

$$\{q \in \mathbb{Q} : q \leq 0 \vee q^2 < 2\},$$

which we have showed to be a cut in Exercise 3.68. But the point here is to *build* $\mathbb{R}$ without assuming knowledge of it, let alone assuming knowledge of the fact that non-empty bounded subsets have a least upper bound. To the contrary, we want this as a by-product of the construction.

Let us define the set

$$\mathbb{R} = \{C \subset \mathbb{Q} : C \text{ is a cut}\}.$$

Proposition 3.106. *The set* $\mathbb{R}$ *is totally ordered by*

$$C \leq D \iff C \subset D.$$

Proof. That the inclusion order on $\mathbb{P}\mathbb{Q}$ restricts to an order relation on $\mathbb{R} \subset \mathbb{P}\mathbb{Q}$ is clear. Let us verify that it restricts to a total order. Let C and D be cuts. We want to show that $C \subset D$ or $D \subset C$. If $C \not\subset D$, then there is $c \in C \setminus D$. Since $\mathbb{Q}$ is totally ordered and D is closed downward in $\mathbb{Q}$, the number c is an upper bound for D. Indeed, for every $d \in D$, either $d \leq c$ or $c < d$. The latter can be ruled out, for then we would have $c \in D$. Thus $d \leq c$. Since C is downward closed, we conclude that $D \subset C$. □

A key property of the reals is that, in contrast to $\mathbb{Q}$, it is *Dedekind complete*, that is, every non-empty subset bounded above has a least upper bound:

Theorem 3.107. *Any non-empty subset of* $\mathbb{R}$ *that has an upper bound also has a least upper bound, and any non-empty subset of* $\mathbb{R}$ *that has a lower bound also has a greatest lower bound.*

Proof. If $\mathscr{A} \subset \mathbb{R}$ is non-empty and has an upper bound $U \in \mathbb{R}$, then $A \subset U$ (in $\mathbb{Q}$) for all $A \in \mathscr{A}$. Thus, $\bigcup_{A \in \mathscr{A}} A \subset U$ is also a cut, and thus the least upper bound of $\mathscr{A}$ in $\mathbb{R}$. Indeed, $\bigcup_{A \in \mathscr{A}} A$ is a subset of U, hence a proper subset of $\mathbb{Q}$, which is closed downward because each $A \in \mathscr{A}$ is. Moreover, it has no greatest element, for if $a \in \bigcup_{A \in \mathscr{A}} A$, there is $A \in \mathscr{A}$ with $a \in A$, so that there is $b \in A \subset \bigcup_{A \in \mathscr{A}} A$ with $a < b$ because A is a cut.

The existence of greatest lower bounds is shown similarly. □

We can identify $\mathbb{Q}$ with a subset of $\mathbb{R}$ by identifying each $q \in \mathbb{Q}$ with the cut

$$C_q = \{x \in \mathbb{Q} : x < q\}.$$

Cuts of the form C_q for some $q \in \mathbb{Q}$ represent rational numbers, while the other cuts represent irrational numbers ([7]).

[7] Note that the map $f : \mathbb{Q} \to \{C_q : q \in \mathbb{Q}\}$ is a bijection from $\mathbb{Q}$ to a subset of $\mathbb{R}$. In that sense, in light of Remark 3.88, the set $\mathbb{Q}$ is obtained as a subset of $\mathbb{R}$.

We can add two real numbers (that is, two cuts) by

$$C+D=\{c+d : c\in D, d\in D\}.$$

Exercise 3.108. Verify that if C and D are cuts then $C+D$ is also a cut.

Multiplication is more complicated to define and takes examination of various cases, but it can be done, albeit tediously. Subtraction and division can be defined similarly.

From now on, we return to thinking of the real line $\mathbb{R}$ in a more traditional manner (as opposed to the set of cuts in $\mathbb{Q}$), but keeping in mind Theorem 3.107.

Note that if S is a non-empty subset of $\mathbb{R}$ that has a greatest lower bound $\inf S$, and a least upper bound $\sup S$, then

$$\forall \varepsilon > 0\ \exists x \in S\ (x \geq \sup S - \varepsilon) \tag{3.12}$$

and

$$\forall \varepsilon > 0\ \exists x \in S\ (x \leq \inf S + \varepsilon). \tag{3.13}$$

Proof (Proof of (3.12)). Otherwise, there would be $\varepsilon_0 > 0$ such that $x < \sup S - \varepsilon_0$ for all $x \in S$, so that $\sup S - \varepsilon_0$ would be an upper bound of S that is smaller than $\sup S$, in contradiction to the very definition of $\sup S$. Of course, (3.13) is verified similarly. □

Suggested Further Readings

Relations are a very general, hence ubiquitous structure in mathematics. As noticed before, graphs can be seen as particular relations, and Graph Theory is a field in itself. Functions and multi-valued functions are also relations and are everywhere. Equivalence relations and quotient sets appear in all areas of mathematics as a standard construction, and a common problem is to transport an additional mathematical structure (such as a group structure, a vector space structure, a topological structure, etc.) from the original space to one of its quotient sets. In that sense, "further readings" to familiarize yourself with various kinds of relations include virtually all of contemporary mathematical writing. Particular emphasis on multi-valued functions (or relations) is paid in certain areas of Analysis and Optimization, where [1] is a good general source.

Yet a special mention should be made of *order relations*. A great many mathematical constructions and structures can be interpreted in order-theoretic terms, particularly through the lens of *Lattice theory*. This is a good entry point to more advanced mathematics with an elementary approach in terms than you are now familiar with. A foundational text is the classical [4]. Grätzer's series of books [16, 17, 18] provides a comprehensive treatment. [14] and [24] provide more advanced accounts of the order-theoretic approach to Topology and Analysis.

Chapter 4
Cardinality

Picking up where we left in Section 1.7, consider again that for finite sets X and Y, the set X has no more elements than Y if and only if there is a one-to-one map $f : X \to Y$, equivalently, if there is an onto map $g : Y \to X$. Moreover, X and Y have the same number of elements if and only if there is a bijection between X and Y. Now we take these observations seriously to extend the concept of cardinality to potentially infinite sets.

4.1 Infinite Sets

As we now consider infinite sets, hence infinite choices, some results depend on the Axiom of Choice, which, as discussed on page 23, we take for granted. The reader more wary of this axiom should know that Proposition 4.3, Theorem 4.7, Proposition 4.9, and Corollary 4.20 all depend on (AC). Further discussion of this axiom and some of its equivalent forms can be found in Appendix A.5.

Definition 4.1. Two (possibly infinite) sets X and Y have *the same cardinality,* in symbols

$$|X| = |Y|,$$

if there is a bijection from X to Y. We say that the cardinality of $|X|$ is *less than or equal* to that of Y, in symbols

$$|X| \leq |Y|,$$

if there is a one-to-one map from X to Y.

Remark 4.2. Note that the definition of $|X| \leq |Y|$ is guided by the following intuition: a subset A of Y "ought to be" of cardinality not greater than that of Y. Moreover, there is a one-to-one map $f : X \to Y$ if and only if there is a bijection from X to some subset of Y (namely $f[X]$), that is, X has the same cardinality as a subset of Y. Hence our definitions allow to extend Corollary 1.106 (1) to all sets, including infinite once. Corollary 1.106 (2) extends as well:

F. Mynard, *An Introduction to the Language of Mathematics*,
https://doi.org/10.1007/978-3-030-00641-9_4

Proposition 4.3. *If $f : X \to Y$ is onto, then $|Y| \leq |X|$.*

Proof. Because f is onto, for every $y \in Y$, $f^{-1}[y] \neq \emptyset$. By the Axiom of Choice (see Section 1.4), we can pick $g(y) \in f^{-1}[y]$ for every $y \in Y$ to the effect that $g : Y \to X$ is one-to-one. Hence $|Y| \leq |X|$. □

Remark 4.4. Since a bijection has an inverse map which is also a bijection and the identity map is a bijection, the relation "has same cardinality as" defines a symmetric and reflexive "relation" on the class of sets. Because a composite of bijections is a bijection (see Exercise 2.13 (3)), it is also transitive. Because the collection of all sets is not a set (see Example 1.62), this is not quite a relation though, but it acts *like* an equivalence relation on sets. A natural question is whether the relation $\leq$ on cardinalities also acts like an order relation on sets. It does. It is clearly reflexive, and it is transitive because a composite of one-to-one maps is one-to-one (Exercise 2.13 (1)). Antisymmetry is not such an easy thing to see, but is established by Theorem 4.5 below.

Theorem 4.5 (Cantor-Bernstein Theorem). *If $|X| \leq |Y|$ and $|Y| \leq |X|$, then $|X| = |Y|$.*

Proof. Because $|X| \leq |Y|$ there is a one-to-one map $f : X \to Y$ and because $|Y| \leq |X|$, there is a one-to-one map $g : Y \to X$. Using a slight abuse of language, we will denote by g^{-1} the inverse map of the bijection obtained from g by restricting the codomain to $g[Y]$.

Let $\mathbb{P}X$ denote the powerset of X ordered by inclusion, which is a complete lattice by Exercise 3.62. Let $\Phi : \mathbb{P}X \to \mathbb{P}X$ be defined by

$$\Phi(M) = g[Y \setminus f[X \setminus M]].$$

This is an order-preserving function, for in general $A \subset B$ implies $f[A] \subset f[B]$ and the two instances of complementation reverse order twice. By Theorem 3.69, Φ has a fixed point $A = \Phi(A) \subset g[Y]$ so that the function $h : X \to Y$ defined by

$$h(x) = \begin{cases} f(x) & \text{if } x \notin A \\ g^{-1}(x) & \text{if } x \in A \end{cases}$$

is well defined for $A \subset g[Y]$. Moreover, h is one-to-one because if there are $x \in A$ and $t \notin A$ with $f(t) = g^{-1}(x)$ then $x = g(f(t)) \in A$, in contradiction to the fact that $g(f(t)) \in g[f[X \setminus A]]$ because

$$x \in A = g[Y \setminus f[X \setminus A]] \subset X \setminus g[f[X \setminus A]],$$

where the last inclusion follows from the fact that g is one-to-one and Theorem 1.139 (6).

On the other hand, h is also onto. To see this, let $y \in Y$. If $y \in f[X \setminus A] = h[X \setminus A]$, we are done. Otherwise, $y \notin f[X \setminus A]$ and then $g(y) \in A$. Then $x = g(y)$ satisfies $y = g^{-1}(x) = h(x)$.

Thus $h : X \to Y$ is a bijection and $|X| = |Y|$. □

Exercise 4.6. Show that if $X \subset Y$ and there is a one-to-one map $f : Y \to X$ then $|X| = |Y|$.

Moreover, the relation $|X| \leq |Y|$ is not only an order, but a total order. Denoting by

$$|X| < |Y|$$

the fact that $|X| \leq |Y|$ and $|X| \neq |Y|$, we have (see the Appendix A.4 for a proof):

Theorem 4.7 (Trichotomy). *If X and Y are sets, then either $|X| < |Y|$ or $|Y| < |X|$ or $|X| = |Y|$.*

Example 4.8 (Hotel Hilbert). Imagine a grand hotel, a very very grand hotel, where rooms are numbered $1, 2, 3, \ldots$ *ad infinitum!* That is, rooms are indexed by natural numbers. Our hotel has infinitely many rooms and will help us illustrate how our intuition based on finite sets is no longer valid in the context of infinite sets. In particular, we should not count on an analog to Remark 1.104 in the context of infinite sets. In the hotel Hilbert (named after mathematician David Hilbert), people are often moved from room to room and the front desk can provide instructions either to a single room, a select group of rooms, or all rooms at once, through an intercom system.

Imagine now that the hotel is full and a new customer shows up and asks for a room. Can she be accommodated? Your intuition probably tells you "of course not!" Yet, room can easily be made for her: ask each guest to move to the next room and voilà! More specifically, tell guests "if you are in room n, please move over to room $n+1$." Then the guest in room 1 moves over to room 2, the one in room 2 moves over to room 3, and so on. Because we have infinitely many rooms, there is *no last room* (unlike for finite sets, and this is why Remark 1.104 does not apply to infinite sets), and thus every guest finds a new room, while room number 1 has been freed for our new guest to use.

Imagine now that, with your hotel full, a bus of tourists comes along seeking rooms. But it is a very, very big bus, with infinitely many seats, labeled $1, 2, \ldots$ ad infinitum! Can we accommodate them all? With the hotel full, we now need to make room for "as many" ([1]) customers as we have rooms! This is not a problem though: tell each guest "if you are in room n, please move over to room $2n$." So the guest in room 1 moves to room 2, that in room 2 moves to room 4, and so on. Again, because we have infinitely many rooms, each guest finds a new room. But now all guests occupy even numbered rooms, and all rooms labeled by odd numbers have been freed up. Ask now each person on the bus to go to the room labeled $2n-1$ if they were occupying the seat labeled n, and you get all your new clients in!

From the above considerations, one notes in particular that the set E of even natural numbers and the set O of odd numbers satisfy

[1] In the sense that the set of new clients C and the set of rooms R are both indexed by $\mathbb{N}$, so that of course

$$|C| = |R| = |\mathbb{N}|.$$

$$|E| = |O| = |\mathbb{N}|,$$

namely because $f : \mathbb{N} \to E$ and $g : \mathbb{N} \to O$, given by $f(n) = 2n$ and $g(n) = 2n - 1$, respectively, are both bijections. Comparing this with Remark 1.104, we observe a characteristic property of infinite sets:

Proposition 4.9. *A set X is infinite if and only if* $|X| = |Y|$ *for some proper subset Y of X.*

Proof. By Remark 1.104, if X has a proper subset of cardinality $|X|$, then X is not finite, hence infinite.

If X is infinite, it contains a sequence of pairwise distinct elements $\{x_n : n \in \mathbb{N}\}$. Let, for instance, $Y = X \setminus \{x_n : n \in O\}$. This is a proper subset of X and $|X| = |Y|$ because the function $f : X \to Y$ defined by $f(x) = x$ for every $x \in X \setminus \{x_n : n \in \mathbb{N}\}$ and $f(x_n) = x_{2n}$ for every $n \in \mathbb{N}$ is easily seen to be a bijection. □

In this argument, we implicitly use that if X is infinite, then there is a one-to-one map $f : \mathbb{N} \to X$ (the sequence $\{x_n : x \in \mathbb{N}\}$)—a fact that I hope to be evident. In other words,

$$|\mathbb{N}| \leq |X|,$$

whenever X is infinite, that is, $|\mathbb{N}|$ is the smallest infinite cardinality:

Definition 4.10. We denote by $\aleph_0$ the cardinality of $\mathbb{N}$; this is the smallest infinite cardinality, that is,

$$\aleph_0 \leq |X|,$$

whenever X is an infinite set.

On the other hand, X is finite if and only if

$$|X| < \aleph_0,$$

for every finite set is of the cardinality of an initial segment $\{1, \ldots n\}$ of $\mathbb{N}$.

Remark 4.11. As noted in Remark 4.4, the equality of cardinalities acts as an equivalence relation on the class of sets (recall that this is not a set). We can think of a given infinite cardinality (such as $\aleph_0$) as representing the equivalence class of all sets with this cardinality, just like we may think of the number 3 as representing the class of all sets with 3 elements, as we did in Remark 3.91.

4.2 Countable Sets

Definition 4.12. We call a set *countable* if it is finite or of cardinality $\aleph_0$, that is, X is countable if and only if $|X| \leq \aleph_0$.

Of course, subsets of $\mathbb{N}$ are countable, and more generally subsets of countable sets are countable. The range of a sequence is countable. In view of Example 4.8, it is not surprising ([2]) that:

Proposition 4.13. *If*

1. *X is countable and $a \notin X$, then $X \cup \{a\}$ is also countable;*
2. *X and Y are countable sets, then $X \cup Y$ is countable.*

Proof. (1). If X is finite, so is $X \cup \{a\}$. If X is infinite, there is a bijection $f : \mathbb{N} \to X$. Let $g : \mathbb{N} \to X \cup \{a\}$ be defined by $g(1) = a$ and $g(n) = f(n-1)$ if $n \geq 2$. ([3]) The map g is easily seen to be a bijection.

(2). This is obvious if X and Y are both finite. If only one is finite, this follows from (1) by induction on the size of the finite set. Assume now that X and Y are both infinite, so that there are bijections $f : \mathbb{N} \to X$ and $g : \mathbb{N} \to X$. Assume first that X and Y are disjoint. Let now $h : \mathbb{N} \to X \cup Y$ be defined by ([4])

$$h(n) = \begin{cases} f\left(\frac{n}{2}\right) & \text{if } n \text{ is even} \\ g\left(\frac{n+1}{2}\right) & \text{if } n \text{ is odd.} \end{cases}$$

The map h is easily verified to be a bijection.

If now X and Y are not disjoint, let Y' be the image of Y under a bijection f so that $Y' \cap X = \emptyset$. Then $h : X \cup Y \to X \cup Y'$ defined by $h(x) = x$ if $x \in X$ and $h(y) = f(y)$ if $y \in Y \setminus X$ is one-to-one. Hence $|X \cup Y| \leq |X \cup Y'|$ and $|X \cup Y'| = \aleph_0$ by (2). Thus $X \cup Y$ is countable. □

Exercise 4.14. Check that the map g in part (1) of the proof of Proposition 4.13, and the map h defined in the second part of the same proof are both bijections.

Corollary 4.15. *The set $\mathbb{Z}$ of integers is countable.*

Proof. $\mathbb{Z} = \mathbb{Z}_- \cup \{0\} \cup \mathbb{Z}_+$ where $\mathbb{Z}_+ = \mathbb{N}$ and $\mathbb{Z}_-$ is the image of $\mathbb{N}$ under the bijection $m(n) = -n$. Hence this follows from Proposition 4.13. □

---

[2] In fact Proposition 4.13 and its proof simply formalize the considerations of Example 4.8.

[3] Reframing this in the context of Example 4.8, g assigns a room number to each element of $X \cup \{a\}$, where X are the existing guests of the hotel, and a is the new client to accommodate. The map $f : X \to \mathbb{N}$ is the initial assignment of room numbers to guests in X. Taking $g(n) = f(n-1)$ for $n \geq 2$ formalizes the idea that the guest in room 1 moves over to room 2, the one in room 2 moves over to room 3, and so on. This way, room 1 is empty and can be assigned to a, that is, $g(1) = a$.

[4] To interpret Example 4.8 in this light, X represents the set of guests in the hotel, and Y the set of new clients in the bus, in which case X and Y are obviously disjoint. f represents the room number assignment of existing guests, and g the seat number assignment in the bus. The map h is our new room number assignment for all people of $X \cup Y$, that is, those in the hotel and those on the bus. The formula $h(n) = f(\frac{n}{2})$ if n is even is the formalization of the fact that the guest in room $m = \frac{n}{2}$ (recall n is even) is sent to room $2m = n$. The formula $h(n) = g\left(\frac{n+1}{2}\right)$ if n is odd formalizes that we place in the odd numbered room $n = 2k - 1$ the passenger of the bus seated at the place labeled $k = \frac{n+1}{2}$.

Remark 4.16. Alternatively, one may easily check directly that $f:\mathbb{N}\to\mathbb{Z}$ defined by

$$f(n)=\begin{cases}\frac{n}{2} & \text{if } n \text{ is even}\\ \frac{1-n}{2} & \text{if } n \text{ is odd}\end{cases}$$

is a bijection.

Theorem 4.17. *The set $\mathbb{N}\times\mathbb{N}$ is countable. More generally, if X and Y are countable, then $X\times Y$ is countable.*

Proof. Let $f:\mathbb{N}\times\mathbb{N}\to\mathbb{N}$ be defined by

$$f(n,p)=2^{n-1}(2p-1).$$

The map f is one-to-one. To see this, assume $2^{n-1}(2p-1)=2^{r-1}(2k-1)$ for n,p,r,k in $\mathbb{N}$. If $n=r$, then $2p-1=2k-1$ and thus $p=k$. Otherwise $n>r$ or $r<n$, say, $n>r$. Then

$$\frac{2^{n-1}}{2^{r-1}}(2p-1)=2^{n-r}(2p-1)=2k-1,$$

which is not possible for $2^{n-r}(2p-1)$ is even and $2k-1$ is odd. Hence $n=r$ and $p=k$, that is, f is one-to-one ([5]), so that

$$|\mathbb{N}\times\mathbb{N}|\leq\aleph_0.$$

If now X and Y are (infinite) countable, there are bijective maps $g:\mathbb{N}\to X$ and $h:\mathbb{N}\to Y$, and therefore and the map $\ell:\mathbb{N}\times\mathbb{N}\to X\times Y$ defined by

$$\ell(n,p)=(g(n),h(p))$$

is easily seen to be a bijection, so that

$$|X\times Y|=|\mathbb{N}\times\mathbb{N}|=\aleph_0.$$

□

Exercise 4.18. Verify that the function $\ell:\mathbb{N}\times\mathbb{N}\to X\times Y$ defined in the proof of Theorem 4.17 is a bijection.

Corollary 4.19. *The set $\mathbb{Q}$ of rational numbers is countable.*

Proof. In view of the construction (3.10) of $\mathbb{Q}$, there is an onto map from $\mathbb{Z}\times\mathbb{Z}\backslash\{0\}$ to $\mathbb{Q}$ so that

[5] While this is enough to conclude, one may easily verify that f is also onto for if $s\in\mathbb{N}$ then $s=2^k t$ for some integer $k\geq 0$ and some odd integer t whenever s is even (factoring all 2's out), and $s=2^0(2n-1)$ for some $n\in\mathbb{N}$ if s is odd. Hence the inequality is an equality, which could also be observed without checking that f is onto by noting that $|\mathbb{N}|\leq|\mathbb{N}\times\mathbb{N}|$ because $i:\mathbb{N}\to\mathbb{N}\times\mathbb{N}$ defined by $i(n)=(1,n)$ is clearly one-to-one.

$$|\mathbb{Q}| \leq |\mathbb{Z} \times \mathbb{Z} \backslash \{0\}| = \aleph_0.$$

□

Corollary 4.20. *If $\{A_i : i \in \mathbb{N}\}$ is a countable collection of countable sets, then*

$$\bigcup_{i \in \mathbb{N}} A_i$$

is also countable.

Proof. For each $i \in \mathbb{N}$, let $f_i : \mathbb{N} \to A_i$ be onto. Consider the map $g : \mathbb{N} \times \mathbb{N} \to \bigcup_{i \in \mathbb{N}} A_i$ defined by $g(n,p) = f_n(p)$. It is clearly onto for if $x \in \bigcup_{i \in \mathbb{N}} A_i$ there is $n \in \mathbb{N}$ with $x \in A_n$, and thus there is $p \in \mathbb{N}$ with $x = f_n(p)$ because f_n is onto. By Proposition 4.3,

$$\left| \bigcup_{i \in \mathbb{N}} A_i \right| \leq |\mathbb{N} \times \mathbb{N}| = \aleph_0.$$

□

Exercise 4.21. Consider the set $P_{\mathbb{Z}}$ of all polynomials with coefficients in $\mathbb{Z}$, and, for each $n \in \mathbb{N}$, let P_n be the subset of $P_{\mathbb{Z}}$ formed by polynomials of degree at most n. Show that P_n is countable for all $n \in \mathbb{N}$ and deduce that $P_{\mathbb{Z}}$ is countable.

4.3 Cardinality Continuum

Let us first observe that any two bounded intervals of the reals have the same cardinality. For instance,

$$|(0,1)| = |(a,b)|$$

for any $a < b$ in $\mathbb{R}$, because the linear map $f : (0,1) \to (a,b)$ defined by $f(t) = a + (b-a)t$ is a bijection. On the other hand,

$$|(a,b)| = |[a,b)| = |(a,b]| = |[a,b]| \tag{4.1}$$

for $(a,b) \subset [a,b]$ and $[a,b] \subset (a-1,b+1)$ so that

$$|(a,b)| \leq |[a,b]| \leq |(a-1,b+1)| = |(a,b)|.$$

Moreover, the tangent map

$$\tan : \left(-\frac{\pi}{2}, \frac{\pi}{2}\right) \to \mathbb{R}$$

is clearly a bijection, so that unbounded intervals also have the same cardinality. Hence

$$|(0,1)| = |(a,b)| = |[a,b]| = |\mathbb{R}|. \tag{4.2}$$

Remark 4.22. We used Theorem 4.5 to easily show (4.1), but what if we wanted an explicit bijection, say, from $(0,1)$ to $(0,1]$? We know that there is one (hence infinitely many), but how to give one explicitly?

We may, for instance, use a variant of the idea used to make room for one more client in our full grand hotel Hilbert in Example 4.8, in which the shift function $n \mapsto n+1$ defined a bijection from $\mathbb{N} \cup \{0\}$ to $\mathbb{N}$, thus "creating" an "additional" room: pick any infinite sequence $\{x_n\}_{n=1}^{\infty}$ of distinct elements in $(0,1)$ (for instance, $x_n = \frac{1}{2n}$) and let $f : (0,1] \to (0,1)$ be defined by $f(x) = x$ for all $x \in (0,1) \setminus \{x_n : n \in \mathbb{N}\}$, $f(1) = x_1$, and $f(x_n) = x_{n+1}$ for $n \in \mathbb{N}$. The additional point 1 is placed in the "emptied room" x_1 after all elements of the sequence are shifted to the right.

Theorem 4.23. *The real line is uncountable.*

Proof. In view of (4.2), it is enough to show that $(0,1)$ is uncountable, and to that end, it is enough to show that there is no onto map $f : \mathbb{N} \to (0,1)$. If a number has two decimal expansions, they end with an infinite string of 0's, or an infinite string of 9's. In such a case, we always pick the one with an infinite string of 0's to write decimal expansions of the values $f(n)$: let $f(1) = 0.a_{1,1}a_{1,2}a_{1,3}\ldots a_{1,n}\ldots$ where $a_{1,1}$, $a_{1,2}$, ... are digits in $\{0,1,2,\ldots 9\}$, and more generally, let

$$f(k) = 0.a_{k,1}a_{k,2}a_{k,3}\ldots a_{k,n}\ldots$$

The number $r = 0.b_1b_2\ldots b_n\ldots$ where ([6])

$$b_n = \begin{cases} 5 & \text{if } a_{n,n} \neq 5 \\ 2 & \text{if } a_{n,n} = 5 \end{cases}$$

is not in the range of f. Indeed, r differs from $f(1)$ in the first decimal place, from $f(2)$ in the second, from $f(n)$ in the nth decimal place, and so on. Thus f is not onto. □

Definition 4.24. The cardinality of the set of real numbers is called *continuum* and denoted by the symbol $\mathfrak{c}$.

Example 4.25. The set $\mathbb{R} \setminus \mathbb{Q}$ of *irrational* numbers is uncountable, for if it was countable, then its union with the countable set $\mathbb{Q}$, which is $\mathbb{R}$, would be countable as well, by Proposition 4.13 (2). Hence there is a lot "more" irrational numbers than rationals.

A number is called *algebraic* if it is the solution of a polynomial equation with rational coefficients. For instance, $\sqrt{2}$ is irrational (see Proposition 2.44) but it is algebraic, for it is a solution of $x^2 - 2 = 0$. Most numbers you are accustomed to manipulate are algebraic, but you know a few numbers that are not, such as π and the Euler number e. Probably against your intuition, it turns out that there are "few"

[6] Note that there is nothing special about the numbers 2 and 5. We could pick any two digits between 1 and 8 (avoiding the possibilities of infinite strings of 0's or 9's), for all we want is to make sure that $b_n \neq a_{n,n}$ for all n.

algebraic numbers, in the sense that the set of algebraic numbers is countable, so that the set of non-algebraic numbers, also called *transcendental numbers*, is uncountable. In other words, there are a lot more transcendental numbers than algebraic numbers. To see this, we can proceed along the lines of Exercise 4.21: consider that for each degree $n \in \mathbb{N}$, coefficients for a polynomial of degree n with leading coefficient 1 are chosen in $\mathbb{Q}^n$, which is a countable set by Corollary 4.19 and Theorem 4.17. Hence the set P_n of polynomial functions of degree n with rational coefficients is countable, so that the set

$$P = \bigcup_{n \in \mathbb{N}} P_n$$

of all polynomial functions with rational coefficients is also countable, by Corollary 4.20. Each polynomial equation has only finitely many solutions, and there are countably many such equations. Hence there are only countably many algebraic numbers.

Proposition 4.26. *The plane $\mathbb{R}^2$ has cardinality $\mathfrak{c}$.*

Proof. The function $f : \mathbb{R} \to \mathbb{R}^2$ defined by $f(x) = (x, 0)$ is clearly injective, so that $\mathfrak{c} \leq |\mathbb{R}^2|$.

On the other hand, consider the map $g : (0,1)^2 \to (0,1)$ sending $(x,y) \in \mathbb{R}^2$ (where $x = 0.x_1x_2 \ldots x_n \ldots$ and $y = 0.y_1y_2 \ldots y_n \ldots$ as decimals) to the number $g(x,y) = 0.x_1y_1x_2y_2 \ldots x_ny_n \ldots$ obtained by interleaving the digits of x and of y. To make this function well-defined, whenever a real number has two decimal expansions (ending with either an infinite string of zeros, or an infinite string of nines) we always (for both x and y) pick the representation ending with zeros. This function is clearly one-to-one and thus $|(0,1)^2| \leq |(0,1)| = \mathfrak{c}$. Moreover, as there is a bijection $f : (0,1) \to \mathbb{R}$, the function $f \times f : (0,1)^2 \to \mathbb{R}^2$ defined by $(f \times f)(x,y) = (f(x), f(y))$ is easily seen to be a bijection, so that $|\mathbb{R}^2| = |(0,1)^2| \leq \mathfrak{c}$, and the result follows from Theorem 4.5. □

More generally,

$$|\mathbb{R}^n| = \mathfrak{c}$$

for every natural number n.

Here are other sets of cardinality $\mathfrak{c}$:

$$|\mathbb{P}(\mathbb{N})| = \left|\{0,1\}^{\mathbb{N}}\right| = \mathfrak{c}. \tag{4.3}$$

To see this, first note that:

Lemma 4.27. *If $|X| \leq |Y|$, then $|\mathbb{P}X| \leq |\mathbb{P}Y|$. In particular, if X and Y have the same cardinality, then $|\mathbb{P}X| = |\mathbb{P}Y|$.*

Proof. Let $f : X \to Y$ be one-to-one. Then $g : \mathbb{P}X \to \mathbb{P}Y$ defined by $g(A) = f[A]$ is one-to-one. Indeed, if $A \neq B$ are subsets of X, then there is $x \in A \setminus B$ or $x \in B \setminus A$.

Say, $x \in A \setminus B$. Then $f(x) \in f[A]$ but $f(x) \notin f[B]$ for otherwise, there would be $b \in B$ with $f(b) = f(x)$, and thus $x = b \in B$ because f is one-to-one. Thus g is one-to-one.

Hence if $|X| = |Y|$, then $|X| \leq |Y| \leq |X|$ and thus $|\mathbb{P}X| \leq |\mathbb{P}Y| \leq |\mathbb{P}X|$ so that $|\mathbb{P}X| = |\mathbb{P}Y|$ by Theorem 4.5. □

Exercise 4.28. Show that if the map f in the proof of Lemma 4.27 is a bijection, then the map g in that proof is also a bijection.

Proof (Proof of (4.3)). By Proposition 1.114, $|\mathbb{P}(\mathbb{N})| = \left|\{0,1\}^{\mathbb{N}}\right|$. Recall from Section 3.4 that $\mathbb{R}$ can be identified with a subset of $\mathbb{P}(\mathbb{Q})$, namely the set of Dedekind cuts. Hence,

$$\mathfrak{c} = |\mathbb{R}| \leq |\mathbb{P}(\mathbb{Q})| = |\mathbb{P}(\mathbb{N})|$$

because of Lemma 4.27 and Corollary 4.19. On the other hand, the map $f : \mathbb{P}(\mathbb{N}) \to (0,1)$ defined by

$$f(A) = 0.a_1 a_2 \dots a_n \dots \text{ where } a_i = \begin{cases} 2 & \text{if } i \in A \\ 5 & \text{if } i \notin A \end{cases} \text{ for each } i \in \mathbb{N},$$

is a well-defined and one-to-one map ([7]). Thus

$$|\mathbb{P}(\mathbb{N})| \leq |(0,1)| = \mathfrak{c},$$

so that $|\mathbb{P}(\mathbb{N})| = \mathfrak{c}$ by Theorem 4.5. □

In particular, we note that

$$|\mathbb{N}| < |\mathbb{P}(\mathbb{N})|,$$

a fact that we will now generalize.

4.4 Infinitely Many Infinite Cardinalities!

Theorem 4.29 (Cantor). *For every set X,*

$$|X| < |\mathbb{P}X|.$$

Proof. First note that $|X| \leq |\mathbb{P}X|$ for the map $i : X \to \mathbb{P}X$ defined by $i(x) = \{x\}$ is one-to-one. To see that $|X| \neq |\mathbb{P}X|$, it is enough to show that a map $f : X \to \mathbb{P}X$ cannot be onto. To this end, consider the set

$$S = \{x \in X : x \notin f(x)\}.$$

[7] As in the proof of Theorem 4.23, there is nothing special about 2 and 5. We are simply making sure that if $A \neq B$ in $\mathbb{P}(\mathbb{N})$ then $f(A) \neq f(B)$, for these two decimal numbers have unique representations (we cannot have strings of 0's or 9's) and they differ in whatever position that is in one set and not in the other.

The set $S \in \mathbb{P}X$ cannot be in the range of f, for if there was $y \in X$ with $S = f(y)$, then we would have

$$y \in S \iff y \notin S,$$

which is clearly a contradiction. Hence f is not onto. □

This is a remarkable result because now

$$|\mathbb{N}| < |\mathbb{P}(\mathbb{N})| < |\mathbb{P}(\mathbb{P}(\mathbb{N}))| < \ldots < |\mathbb{P}(\ldots(\mathbb{P}(\mathbb{P}(\mathbb{N}))))| < \ldots \tag{4.4}$$

defines a strictly increasing infinite sequence of infinite cardinalities!

In view of Proposition 1.114, $|\mathbb{P}X| = \left|\{0,1\}^X\right|$ leading to the following notation: if $|X| = \kappa$ then

$$2^\kappa = |\mathbb{P}X| > \kappa.$$

Hence (4.4) rephrases in those terms as

$$\aleph_0 < 2^{\aleph_0} = \mathfrak{c} < 2^{\mathfrak{c}} < 2^{2^{\mathfrak{c}}} \ldots < 2^{2^{\cdot^{\cdot^{\cdot^{2^{\mathfrak{c}}}}}}} < \ldots$$

4.5 Continuum Hypothesis and the Surprisingly Complex Nature of Truth in Mathematics

We have seen that $\aleph_0$ is the smallest infinite cardinality. It turns out that there is the smallest cardinality greater than $\aleph_0$ denoted $\aleph_1$, and the smallest cardinality $\aleph_2$ that is greater than $\aleph_1$, and so on ([8]). Since $\mathfrak{c} = 2^{\aleph_0} > \aleph_0$, we have by definition

$$\aleph_1 \leq \mathfrak{c}.$$

A natural question is whether there is in fact equality or if $\aleph_1 < \mathfrak{c}$. Cantor conjectured that

$$\aleph_1 = \mathfrak{c} \tag{CH}$$

and spent considerable efforts trying to prove what he coined the *Continuum Hypothesis* abbreviated (CH). Because (CH) would have many important consequences in various branches of mathematics, this was the first in the list of 23 problems presented by David Hilbert, one of the most influential mathematicians of his time, at the International Congress of Mathematicians in Paris in 1900. These problems, some of which remain open, have had a tremendous influence on mathematical research throughout the 20th century.

Completing earlier work of Kurt Gödel from 1940, Paul Cohen proved in 1963 that (CH) is independent of the usual axioms ZFC of set-theory, which earned him

[8] We will take this as a fact, even though we do not fully justify it, even in the Appendix, as we avoid introducing ordinal numbers.

the 1966 Fields Medal, the highest recognition in mathematics ([9]). That means that you may add (CH) to ZFC as an additional axiom if you choose to, or, alternatively, you can add the negation of (CH) as an additional axiom. These two choices produce two mathematical realities that largely overlap, but some features of the mathematical landscape are drastically different depending on whether you take (CH) or its negation.

(CH) being tightly tied to other conjectures in Set-theoretic Topology, Analysis and Measure Theory, this proved other important statements to be independent of the usual axioms. Even though the famous 1931 *incompleteness theorems* of Kurt Gödel had shown that any axiomatic system containing basic arithmetic can formulate statements independent of those axioms (and that no such axiomatic system can prove its own consistency ([10])) such situations didn't really happen in the daily work of mathematicians. That statements of mathematical significance can be independent of the usual axioms of set-theory came as somewhat of a shock.

Additional Exercises for Chapter 4

Exercise 4.30. Prove or disprove:

1. if there is a one-to-one map $f : X \to Y$ and an onto map $g : X \to Y$, then there is a bijection $h : X \to Y$.
2. if X is a set and Y is an uncountable set and there is an onto map $f : X \to Y$, then X is uncountable.
3. If X is uncountable, then $|X| = |\mathbb{R}|$.
4. Every infinite set is the subset of a countable infinite set.
5. The set of sequences on $\mathbb{N}$ is countable.

Exercise 4.31. Let $f : X \to Y$, $g : Y \to Z$ and $h : Z \to X$ be three one-to-one functions. Either prove that $|X| = |Y| = |Z|$ or give a counterexample.

Exercise 4.32. Give an example (or explain why it is not possible) of a countable collection of finite sets

1. whose union is finite;
2. whose union is countable and infinite;
3. that is pairwise disjoint and whose union is finite.

Exercise 4.33. For each of the following sets, explain why it is countable, or why it is uncountable:

[9] Specifically, Gödel had proved that (CH) cannot be disproved in ZFC, and Cohen proved that (CH) cannot be proved in ZFC, thus completing the proof of independence from ZFC.

[10] A fact that one can put that way: doing mathematics involves an act of faith. You need to "believe" in your axiomatic system, because this axiomatic system cannot prove its own consistency, that is, that it is free of contradictions, even though you can prove the consistency of an axiomatic system "from outside."

1. $\mathbb{Q}^{10}$;
2. $\mathbb{P}(\mathbb{Q})$;
3. The set of finite subsets of $\mathbb{N}$;
4. $\{0,1\} \times (\mathbb{R} \setminus \mathbb{Q})$;
5. $\mathbb{Q} \times \mathbb{Z}$;
6. $A = \{(x,y) \in \mathbb{R}^2 : x \in \mathbb{Q}, y \in \mathbb{Q} \setminus \mathbb{Z}\}$;
7. $\bigcup_{n \in \mathbb{N}} \left\{ \frac{n}{2^k} : k \in \mathbb{N} \right\}$;
8. $\mathbb{R}^2 \setminus A$ where A is as in 6;
9. The closed unit disk $\{(x,y) \in \mathbb{R}^2 : x^2 + y^2 \leq 1\}$ of the plane.

Exercise 4.34. Give an explicit bijection between:

1. $(0,\infty)$ and $\mathbb{R}$;
2. $\mathbb{Z}$ and $\{x \in \mathbb{R} : \sin x = -1\}$;
3. The sets $\{3k : k \in \mathbb{Z}\}$ and $\{5k : k \in \mathbb{Z}\}$;
4. The intervals $(0,1)$ and $[2,4]$.

Exercise 4.35. Arrange the following cardinality in increasing order, using equality whenever applicable:

$$|\emptyset|, |\mathbb{R}|, |\mathbb{P}(\mathbb{R})|, |\{\emptyset, \{\emptyset\}\}|, |\mathbb{N}|, |\mathbb{P}(\mathbb{P}(\mathbb{R}))|, |\mathbb{Q}|, |\mathbb{R} \setminus \mathbb{Q}|, |\mathbb{P}(\mathbb{P}(\mathbb{Q}))|, |\{\emptyset\}|, |\mathbb{P}(\mathbb{Q})|.$$

Exercise 4.36. What is the cardinality of the set $\mathbb{Q}^{\mathbb{N}}$ of sequences of rational numbers?

Exercise 4.37. Is there a greatest infinite cardinality? Why or why not.

Exercise 4.38. Find an example of a map as below, or explain why it is not possible:

1. A one-to-one map $f : \mathbb{R} \to \mathbb{Q}$;
2. A one-to-one map $g : \mathbb{P}(\mathbb{N}) \to \mathbb{N}$;
3. An onto map $h : \mathbb{P}(\mathbb{Q}) \to \mathbb{R}$;
4. A one-to-one map $p : \mathbb{Q} \times \mathbb{Q} \to \mathbb{N}$;
5. An onto map $q : \mathbb{N} \to \mathbb{Q} \times \mathbb{Q}$;
6. An onto map $r : \mathbb{R} \to \mathbb{P}(\mathbb{P}(\mathbb{N}))$.

Suggested Further Readings

Some of the readings suggested at the end of Chapter 1 should be revisited now, such as [3], [7], [22].

[15] is an interesting and relatively accessible article showing that Theorem 4.7 is equivalent to the Axiom of Choice, and that a generalized form of the Continuum Hypothesis implies the Axiom of Choice. The reader wanting to learn more on the Continuum Hypothesis and on Logic may want to read the classic [9] of Paul Cohen.

Appendix A
Complements

A.1 Inclusion-Exclusion and Number of onto Maps

To complete the picture emerging from Proposition 1.112 in Section 1.7, we turn to the number of onto maps between finite sets. To this end, we start with an important counting principle, known as the *principle of inclusion and exclusion*:

Theorem A.1 (Principle of Inclusion-Exclusion). *If $\{S_i : i \in I\}$ is a family of n subsets (that is $|I| = n$) of a fixed finite set F, consider for each subset J of I the set*

$$S_J = \bigcap_{i \in J} S_i$$

with the convention that $S_\emptyset = F$. The number of elements of F which lie in none of the sets S_i is

$$|F \setminus \bigcup_{i \in I} S_i| = \sum_{J \in \mathbb{P}(I)} (-1)^{|J|} |S_J|.$$

Proof. Because $(-1)^{|J|}|S_J|$ corresponds to a contribution of $(-1)^{|J|}$ for each $x \in S_J$, we can focus on the contribution of a given $x \in F$, which will be the sum of the $(-1)^{|J|}$ over the subsets J of I such that $x \in S_J$, to the effect that

$$\sum_{J \in \mathbb{P}(I)} (-1)^{|J|} |S_J| = \sum_{x \in F} \sum_{J \in \mathbb{P}(I): x \in S_J} (-1)^{|J|}. \tag{A.1}$$

If $x \in F \setminus \bigcup_{i \in I} S_i$, then the only J for which $x \in S_J$ is $J = \emptyset$. Hence, for such an x,

$$\sum_{J \in \mathbb{P}(I): x \in S_J} (-1)^{|J|} = (-1)^0 = 1.$$

If now $x \in \bigcup_{i \in I} S_i$, let $J_x = \{i \in I : x \in S_i\}$ and let $j = |J_x|$. Then subsets J of I with $x \in S_J$ are exactly subsets of J_x, so that

F. Mynard, *An Introduction to the Language of Mathematics*,
https://doi.org/10.1007/978-3-030-00641-9

$$\sum_{J\in\mathbb{P}(I):x\in S_J}(-1)^{|J|}=\sum_{J\subset J_x}(-1)^{|J|}=\sum_{k=0}^{j}\binom{j}{k}(-1)^k,$$

because there are $\binom{j}{k}$ subsets of J_x with k elements. In view of Theorem 1.120,

$$\sum_{k=0}^{j}\binom{j}{k}(-1)^k=\sum_{k=0}^{j}\binom{j}{k}(1)^{j-k}(-1)^k=(1-1)^j=0.$$

Hence, in (A.1), each $x\in F\setminus\bigcup_{i\in I}S_i$ contributes 1, and each $x\in\bigcup_{i\in I}S_i$ contributes 0. In other words,

$$|F\setminus\bigcup_{i\in I}S_i|=\sum_{x\in F}\sum_{J\in\mathbb{P}(I):x\in S_J}(-1)^{|J|}=\sum_{J\in\mathbb{P}(I)}(-1)^{|J|}|S_J|.$$

□

Corollary A.2. *If $\{S_i : i\in I\}$ is a finite collection of finite sets, then*

$$|\bigcup_{i\in I}S_i|=\sum_{J\in\mathbb{P}(I)\setminus\{\emptyset\}}(-1)^{|J|-1}|S_J|. \tag{A.2}$$

Proof. Let $F=\bigcup_{i\in I}S_i$ and apply Theorem A.1 to the effect that

$$0=\sum_{J\in\mathbb{P}(I)}(-1)^{|J|}|S_J|.$$

When $J=\emptyset$, $(-1)^{|J|}|S_J|=|F|$, so that

$$\sum_{J\in\mathbb{P}(I)}(-1)^{|J|}|S_J|=|F|+\sum_{J\in\mathbb{P}(I)\setminus\{\emptyset\}}(-1)^{|J|}|S_J|=0,$$

that is,

$$|F|=\sum_{J\in\mathbb{P}(I)\setminus\{\emptyset\}}(-1)^{|J|-1}|S_J|.$$

□

Example A.3. Let us spell out Corollary A.2 in the case of three sets A, B, and C. Then I consists of 3 indices and the possible sets S_J for $J\neq\emptyset$ are A, B, and C when $|J|=1$, $A\cap B$, $A\cap C$ and $B\cap C$ when $|J|=2$ and $A\cap B\cap C$ when $|J|=3$. Thus (A.2) becomes

$$|A\cup B\cup C|=|A|+|B|+|C|-|A\cap B|-|A\cap C|-|B\cap C|+|A\cap B\cap C|.$$

This is easily understood from a picture:

In $|A|+|B|+|C|$, elements of $B\cap C$, $A\cap B$, and $A\cap C$ that are not in $A\cap B\cap C$ are counted twice, and those of $A\cap B\cap C$ are counted three times. In removing $|A\cap B|+|A\cap C|+|B\cap C|$, we remove $|A\cap B\cap C|$ three times, so it has to be added back.

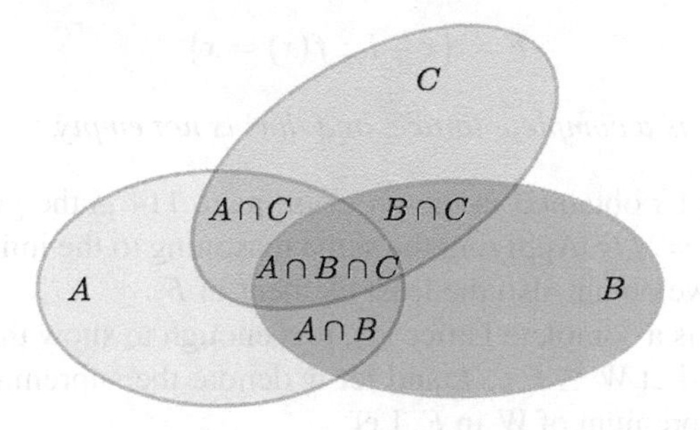

Exercise A.4. How many integers between 1 and 100 are not divisible by 2, 3, or 5?

Exercise A.5. In a high school class of 40 students, 18 like Mathematics, 16 like English, and 12 like PE. On the other hand, 7 like both Math and English, 5 like both Math and PE, and 3 like both English and PE. Finally, 2 like all 3 subjects. How many students do not like any of these 3 subjects?

Corollary A.6. *Let $|X| = n$, $|Y| = k$ and $n \geq k$. The number of onto maps $f : X \to Y$ is*

$$\sum_{j=0}^{k}(-1)^j\binom{k}{j}(k-j)^n.$$

Proof. Let F denote the set Y^X of functions from X to Y. For each $y \in Y$, let S_y be the set of functions that do not take the value y. Obviously, f is onto if and only if $f \in Y^X \setminus \bigcup_{y\in Y} S_y$. Given $J \subset Y$, the set S_J consists of all functions that take no value in J, that is, S_J is the set $(Y \setminus J)^X$ and thus, in view of Theorem 1.115,

$$|S_J| = (k-j)^n,$$

where $j = |J|$. Moreover, there are $\binom{k}{j}$ subsets J of Y of cardinality j, each of them contributing $(-1)^j(k-j)^n$ in the inclusion-exclusion sum of Theorem A.1, so that

$$\begin{aligned}|Y^X \setminus \bigcup_{y\in Y} S_y| &= \sum_{J\in\mathbb{P}(Y)} (-1)^{|J|}|S_J| \\ &= \sum(-1)^j\binom{k}{j}(k-j)^n.\end{aligned}$$

□

A.2 Knaster-Tarski Fixed Point Theorem

Theorem A.7 (Knaster-Tarski Fixed Point Theorem). *If $(L, \leq)$ is a complete lattice and $f : L \to L$ is order-preserving, then the set*

$$F = \{x \in L : f(x) = x\}$$

of fixed points of f is a complete lattice, and thus is not empty.

Proof. The element s obtained in the proof on page 114 is the greatest element of F, for $F \subset D$ and $s = \bigvee D$. Applying the same reasoning to the infimum of $U = \{x \in L : x \geq f(x)\}$ ([1]), we obtain also the least element of F.

To show that F is a complete lattice ([2]), it is enough to show that its every subset have a supremum. Let $W \subset F \subset L$ and let w denote the supremum in L of W. We want to find the supremum of W in F. Let

$$[w, \top] = \{x \in L : w \leq x \leq \top\},$$

where $\top$ denotes the greatest element of L.

We claim that

$$f([w, \top]) \subset [w, \top], \tag{A.3}$$

which is sufficient to conclude, for then f restricts to an order-preserving map of the complete lattice $[w, \top]$ (for the order induced by that of L) and thus has a least fixed point, say m. Then m is the least element of F that is greater than w, that is, m is the supremum of W in F.

To see the claim, note that if $x \in W \subset F$ then $x \leq w$ so that

$$x = f(x) \leq f(w)$$

because f is order-preserving and x is a fixed point of f. Therefore, $f(w)$ is an upper bound of W and thus $w \leq f(w)$. If now $y \in [w, \top]$, that is, $w \leq y$, then

$$w \leq f(w) \leq f(y),$$

that is, $f(y) \in [w, \top]$, which completes the proof. □

A.3 Induction Revisited and Well-Ordered Sets

Let us revisit Induction, presented somewhat naively in Section 2.6, in a more formal fashion.

What we invoke when proving a result by induction is the following property of $\mathbb{N}$ that we will call PMI (for *principle of mathematical induction*): If T is a subset of $\mathbb{N}$ with $1 \in \mathbb{N}$ and

$$\forall n \in \mathbb{N}\ (n \in T \Longrightarrow n+1 \in T) \tag{A.4}$$

then $T = \mathbb{N}$.

---

[1] Or simply applying the previous result to D for the reverse order (which is also a complete lattice).

[2] This second part of the proof is a little bit more sophisticated, but is, again, not really needed if we only want to know that $F \neq \emptyset$.

In practice, in proving a statement of the form

$$\forall n \in \mathbb{N}\, P(n)$$

we consider the truth set T of $P(n)$ and verify that T satisfy the conditions in the PMI to conclude that $T = \mathbb{N}$, that is, that $P(n)$ is true for all $n \in \mathbb{N}$.

On the other hand, when proving something by strong induction, we invoke the following property of $\mathbb{N}$ that we will call PSI (for *principle of strong induction*, often called *principle of complete induction*, e.g., [27]): If $T \subset \mathbb{N}$ satisfies

$$\forall n\ (\{1,2,\ldots n-1\} \subset T \Longrightarrow n \in T), \tag{A.5}$$

then $T = \mathbb{N}$.

Remark A.8. You may wonder why this form of induction seemingly does not include a base case in which we show that $1 \in T$. This is "hidden" in (A.5): the instance for $n = 1$ is

$$\emptyset \subset T \Longrightarrow 1 \in T,$$

whose verification amounts to showing that $1 \in T$.

While we have mentioned that induction and strong induction are equivalent, we didn't formally prove that they are. We will do that now, adding a third equivalent form, that we will call WOP for *well-ordering principle*: every non-empty subset of $\mathbb{N}$ has a smallest element.

Theorem A.9. *PMI, PSI, and WOP are equivalent.*

Proof. PMI $\Longrightarrow$ PSI. Assume PMI and let $T \subset \mathbb{N}$ satisfy (A.5). We show by PMI that $\{1,\ldots n\} \subset T$ for all $n \in \mathbb{N}$, so that $T = \mathbb{N}$, proving PSI. The case $n = 1$ in (A.5) yields $1 \in T$ (see Remark A.8), hence $\{1\} \subset T$. If $\{1,\ldots n\} \subset T$, then (A.5) yields $n+1 \in T$, By PMI, we conclude that $\mathbb{N} = T$.

PSI $\Longrightarrow$ WOP. Assume PSI and let A be a non-empty subset of $\mathbb{N}$, and suppose by way of contradiction that A has no smallest element. Let $T = \mathbb{N} \setminus A$. We show that T satisfies (A.5) so that $T = \mathbb{N}$ by PSI, in contradiction to $A \neq \emptyset$. To this end, let $n \in \mathbb{N}$ with $\{1,\ldots n-1\} \subset T$. Hence none of the numbers 1 through $n-1$ belongs to A. Hence $n \notin A$ for otherwise n would be the smallest element of A. Thus $n \in T$. Therefore T satisfies (A.5) and the result follows.

WOP $\Longrightarrow$ PMI. Assume WOP and that T is a subset of $\mathbb{N}$ with $1 \in T$ and (A.4). We need to show that $T = \mathbb{N}$. Let $A = \mathbb{N} \backslash T$. If $A = \emptyset$, we are done. Otherwise, WOP applies to the effect that A has a smallest element n. Moreover, $n \geq 2$ for $1 \in T$. Thus $n-1 \geq 1$ belongs to T because $n-1 < n$ and n is the smallest element of A. By (A.4), $n \in T$, which contradicts $n \in A$. Thus $A = \emptyset$ and $T = \mathbb{N}$. □

WOP finds a natural order-theoretic generalization, which allows for a more abstract version of induction.

Definition A.10. A poset $(X, \leq)$ is a *well-ordered set* if every non-empty subset has a least element.

Example A.11. The set of natural numbers with its usual order is a well-ordered set. It is the "simplest" infinite well-ordered set. A finite totally ordered set is also well-ordered.

Proposition A.12. *A well-ordered set is totally ordered.*

Proof. Let $x \neq y \in X$. Since $\{x,y\} \subset X$ is non-empty and X is well-ordered, $\{x,y\}$ has a smallest element. Hence, either $x \leq y$ or $y \leq x$. □

On the other hand, there are totally ordered sets that are not well-ordered. For instance, $\mathbb{R}$ and its subsets $\mathbb{Z}$ and $\mathbb{Q}$ with their usual order are totally ordered, but not well-ordered.

As we have seen, it is the well-ordered nature of $(\mathbb{N}, \leq)$ that permits proofs by induction (because WOP, PMI, and PSI are equivalent), and thus, induction can be generalized to other well-ordered sets as we will see shortly. As usual, if $\leq$ is an order relation, we denote by $<$ the associated strict order defined by $x < y$ if $x \leq y$ and $x \neq y$.

Theorem A.13 (General Principle of Induction). *Let $(X, \leq)$ be a well-ordered set, and let $P(x)$ be a predicate on X. Suppose that*

$$\forall x \in X \left((\forall t < x \; P(t)) \Longrightarrow P(x) \right). \tag{A.6}$$

Then X is the truth set of P (that is, $\forall x \; P(x)$).

Proof. Let Y be the truth set of P in X. If $Y \neq X$, then $X \setminus Y \neq \emptyset$. Since X is well-ordered, $X \setminus Y$ has a least element, say x. For every $t < x$, t cannot be in $X \setminus Y$, hence belongs to Y. Hence, $P(t)$ for all $t < x$ and thus $P(x)$ is true by (A.6), that is, $x \in Y$, contrary to the assumption that x is a (least) element of $X \setminus Y$. Thus $X = Y$. □

In other words, any well-ordered set yields a version of the principle of strong induction, which we call *general principle of induction* and is the natural generalization of induction from $\mathbb{N}$ to a general well-ordered set. Indeed, the form PMI is specific to $\mathbb{N}$.

Remark A.14. Note that, like noted for PSI on $\mathbb{N}$, the general principle of induction formally does not include a base case. It is because it is "built-in" in the formulation (A.6). Indeed, if x is the least element $\perp$ of X, there is no $t \in X$ with $t < x$, so that the premise $(\forall t < x \, P(t))$ is vacuously true (all $t < x$ satisfy $P(t)$ because there is no such element!). Hence $P(\perp)$ is true. In other words, checking (A.6) for $x = \perp$ consists in checking $P(\perp)$.

A.4 Well-Order and Trichotomy

We develop here enough basic material on well-order sets to prove Theorem 4.7.

Recall that if X and Y are ordered sets and $f : X \to Y$ is a map, then f is *order-preserving* or *increasing* if

$$x_0 \leq x_1 \Longrightarrow f(x_0) \leq f(x_1),$$

and *strictly increasing* if

$$x_0 < x_1 \Longrightarrow f(x_0) < f(x_1).$$

Proposition A.15. *If X is well-ordered and $f : X \to X$ is strictly increasing, then*

$$x \leq f(x),$$

for every $x \in X$.

Proof. Otherwise,

$$A = \{x \in X : x > f(x)\} \neq \emptyset.$$

As X is well-ordered, there is the smallest element x_0 of A. Therefore $x_1 = f(x_0) < x_0$ and, since f is strictly increasing, $f(x_1) < f(x_0) = x_1$, that is, $x_1 \in A$ and $x_1 < x_0$, which is a contradiction. □

If f is increasing and bijective and if the inverse map f^{-1} is increasing, then f is called an *order isomorphism.* Two ordered sets X, Y are said to be *(order) isomorphic*, in symbols $X \simeq Y$, if there exists an order isomorphism between them.

Proposition A.16. *If X is a well-ordered set and $f : X \to X$ is an order isomorphism, then f is the identity: $f(x) = x$ for each $x \in X$.*

Proof. If f is an isomorphism, then by Proposition A.15, $x \leq f(x)$ and, on the other hand, $y \leq f^{-1}(y)$ for each $y \in X$, hence $x \leq f(x) \leq f^{-1}(f(x)) = x$. □

Proposition A.17. *If X and Y are well-ordered and isomorphic, then there exists a unique isomorphism.*

Proof. If $f : X \to Y$ and $g : X \to Y$ are isomorphisms, then $g^{-1} \circ f : X \to X$ is an isomorphism, hence by Proposition A.16, $g^{-1} \circ f = i_X$ and thus $g = f$. □

Proposition A.18. *If X, Y are totally ordered* ([3]) *and $f : X \to Y$ is a strictly increasing bijection, then f is an isomorphism.*

Proof. Let us show that $f^{-1} : Y \to X$ is strictly increasing by showing the contrapositive, that is, that if $f^{-1}(y_0) \nless f^{-1}(y_1)$ then $y_0 \geq y_1$. If $f^{-1}(y_0) \nless f^{-1}(y_1)$ then $f^{-1}(y_0) \geq f^{-1}(y_1)$, because the order of X is total. Hence

$$y_0 = f\left(f^{-1}(y_0)\right) \geq f\left(f^{-1}(y_1)\right) = y_1,$$

because f is increasing. □

[3] This proposition is no longer valid if the order of the domain is not total.

If W is well-ordered, then for each $x \in W$, the set

$$W_{\downarrow x} = \{w \in W : w < x\}$$

is called the *initial segment* of W corresponding to x.

Proposition A.19. *A well-ordered set is not isomorphic to any of its initial segments.*

Proof. If W is well-ordered, $x \in W$ and $f : W \to W_{\downarrow x}$ is strictly increasing, then $f(x) \in W_{\downarrow x}$, hence $f(x) < x$, in contradiction with Proposition A.15. □

Theorem A.20 (Trichotomy for Well-Ordered Sets). *If X, Y are well-ordered, then exactly one of the following situations holds:*

1. $X \simeq Y$,
2. *there exists* $y_0 \in Y$ *such that* $X \simeq Y_{\downarrow y_0}$,
3. *there exists* $x_0 \in X$ *such that* $Y \simeq X_{\downarrow x_0}$.

Proof. Let

$$F = \{(x,y) \in X \times Y : X_{\downarrow x} \simeq Y_{\downarrow y}\}.$$

If $y_0, y_1 \in F(x)$ then $Y_{\downarrow y_0} \simeq X_{\downarrow x} \simeq Y_{\downarrow y_1}$, hence $y_0 = y_1$, that is, $F(x)$ is a singleton, and F is a functional relation with domain $F^{-1}(Y)$. For the same reason, F^{-1} is a functional relation with domain $F(X)$, so that there is a bijective map $f : F^{-1}(Y) \to F(X)$ such that $F(x) = \{f(x)\}$ for each $x \in F^{-1}(Y)$.

Let $(x_0, y_0) \in F$ and let $h : X_{\downarrow x_0} \to Y_{\downarrow y_0}$ be an isomorphism. Because h is an order isomorphism, $X_{\downarrow x} \simeq Y_{\downarrow h(x)}$ for each $x < x_0$, hence $(x, h(x)) \in F$, that is, $f(x) = h(x)$ and $h(x) < y_0$. This shows that f is strictly increasing and that if $x < x_0$ and $x_0 \in F^{-1}(Y)$, then $x \in F^{-1}(Y)$. Symmetrically, if $y < y_0 \in F(X)$, then $y \in F(X)$.

If $F(X) = Y$ and $F^{-1}(Y) = X$, then $X \simeq Y$ by Proposition A.18, because f is strictly increasing and bijective.

If $F(X) \neq Y$, then let y_0 denote the smallest element of $Y \setminus F(X)$, so that $F(X) = Y_{\downarrow y_0}$. It follows that $F^{-1}(Y) = X$, for otherwise there would exist the smallest element x_0 of $X \setminus F^{-1}(Y)$, so that $F^{-1}(Y) = X_{\downarrow x_0}$. Then $f : X_{\downarrow x_0} \to Y_{\downarrow y_0}$ is an isomorphism, so that $(x_0, y_0) \in F$ in contradiction with $x_0 \in X \setminus F^{-1}(Y)$. Therefore $X \simeq Y_{\downarrow y_0}$. Finally if $F^{-1}(Y) \neq X$, then $X_{\downarrow x_0} \simeq Y$, where x_0 is the smallest element of $X \setminus F^{-1}(Y)$, by a similar argument. □

As we have seen in Theorem A.13, having a set well-ordered opens the possibility to perform induction on its elements. As induction is a powerful tool, it is natural to ask if any set can be well-ordered, that is, if on any set we can find a total order relation for which every non-empty subset has a least element. This is indeed a fact in ZFC, which turns out to be equivalent to the Axiom of Choice (AC) introduced on page 23:

Theorem A.21 (Well-Ordering Theorem (WO)). *Every set can be well-ordered.*

Therefore Theorem 4.7 follows immediately from Theorem A.20 and Theorem A.21:

Proof (Proof of Theorem 4.7). Let X and Y be sets. By the Well-ordering Theorem, we can find order relations on X and Y turning them into well-ordered sets. By Theorem A.20, either $X \simeq Y$ and thus $|X| = |Y|$, or $X \simeq Y_{\downarrow y_0}$ for some $y_0 \in Y$ and then

$$|X| = |Y_{\downarrow y_0}| < |Y|$$

by Proposition A.19, or $Y \simeq X_{\downarrow x_0}$ for some $x_0 \in X$ and

$$|Y| = |X_{\downarrow x_0}| < |X|.$$

□

A.5 More on the Axiom of Choice

Here is another often useful order-theoretic reformulation of (AC) known as *Kuratowski-Zorn Lemma* or sometimes as *Zorn's Lemma*:

Lemma A.22 (Kuratowski-Zorn Lemma (ZL)). *If X is a poset with the property that every totally ordered subset of X has an upper bound (in X), then for each $x \in X$, there is a maximal element m of X with $x \leq m$.*

Theorem A.23. *The following are equivalent:*

1. *the Axiom of Choice (AC);*
2. *the Well-ordering Theorem (WO);*
3. *Zorn's Lemma (ZL).*

Proof. $(2) \Longrightarrow (1)$: if X is a set of non-empty sets, (WO) allows to consider the least element $f(A)$ of each $A \in X$, yielding a choice function (see Footnote 12 on page 23).

$(1) \Longrightarrow (3)$:

Assume (AC), and assume by contradiction that X is a poset in which every totally ordered subset has an upper bound, with $x_0 \in X$ and s is not maximal whenever $s \in X_0 := \{x \in X : x_0 \leq x\}$. For every totally ordered subset T of X_0 there is $f(T)$ satisfying $t < f(T)$ for all $t \in T$, because T has an upper bound m_T, which is not maximal, so that we can pick (with (AC)!) $f(T) > m_T$. Let $\text{well}(X_0)$ denote the set of well-ordered subsets of X_0. Of course, $\text{well}(X_0) \neq \emptyset$ for the empty set and singletons are members. We call $W \in \text{well}(X_0)$ *f-inductive* if

$$\forall x \in W\ x = f(\{y \in W : y < x\}).$$

If Y and Z are elements of $\text{well}(X_0)$, in view of Theorem A.20, then $Y \simeq Z$ or one is isomorphic to an initial segment of the other. In each case, the isomorphism is unique by Proposition A.17. If moreover Y and Z are f-inductive, then this isomorphism is an identity map. Indeed, if, for instance, $Y \simeq Z_{\downarrow z_0}$ for some $z_0 \in Z$, and $g : Y \to Z_{\downarrow z_0}$ is the isomorphism, then

$$\begin{aligned} g(y) &= f\left(\left\{s \in Z_{\downarrow y_0} : s < g(y)\right\}\right). \\ &= f\left(\left\{s \in Z_{\downarrow z_0} : g^{-1}(s) < y\right\}\right) \\ &= f(\{t \in Y : t < y\}) = y. \end{aligned}$$

Hence the collection $\mathscr{A}$ of f-inductive well-ordered subsets of X_0 is totally ordered by inclusion of initial segment. Moreover, $\bigcup \mathscr{A} \in \mathscr{A}$. Indeed, $\bigcup \mathscr{A}$ is well-ordered for $\leq$: first note that $\bigcup \mathscr{A}$ is totally ordered, because for any two different elements, one is an initial segment of the other. Second, note that if $\emptyset \neq S \subset \bigcup \mathscr{A}$, then there is $A \in \mathscr{A}$ with $S \cap A \neq \emptyset$ so that there is a minimal element m of $S \cap A$, since A is well-ordered. If $t \in S$ with $t \leq m$, then $t \in A$ because A is an initial segment, so that $t = m$. Hence m is minimal in S. As the order is total in $\bigcup \mathscr{A}$, we conclude that m is the smallest element of S. Moreover, $\bigcup \mathscr{A}$ is also f-inductive for if $x \in \bigcup \mathscr{A}$ there is $A \in \mathscr{A}$ with $x \in A$ so that $x = f(\{t \in A : t < x\})$ and

$$\{t \in A : t < x\} = \left\{t \in \bigcup \mathscr{A} : t < x\right\}$$

because if $t \in \bigcup \mathscr{A}$ with $t < x$, then $t \in A$ because A is an initial segment.

Since $\bigcup \mathscr{A} \in \mathscr{A}$, the set $\bigcup \mathscr{A}$ is the largest f-inductive set. Yet, $\bigcup \mathscr{A} \cup \{f(\bigcup \mathscr{A})\}$ is a larger f-inductive set, yielding a contradiction.

$(3) \Longrightarrow (2)$: Let X be a set and let Y be the set of pairs (A, R) where $A \in \mathbb{P}X$, and R is a well-order on A. The set Y is non-empty, for the empty set and singletons can be well-ordered. Order Y by

$$(A, R) \leq (B, S)$$

if (A, R) is an initial segment of B for S (it is a simple verification that this is indeed an order on Y). With an argument similar to that used to show that $\bigcup \mathscr{A}$ is well-ordered in $(1) \Longrightarrow (3)$, it is easily seen that whenever $\mathscr{C}$ is a chain in Y, the set $\bigcup \mathscr{C}$ is well-ordered. By Zorn's Lemma, Y has maximal elements. Let (A, R) be maximal in Y. Then $A \subset X$. If $A \neq X$, there is $x \in X \setminus A$. Let $B = A \cup \{x\}$ and order B by

$$b_1 \leq b_2 \iff \begin{cases} b_1 R b_2 & \text{if } b_1, b_2 \in A \text{ or} \\ \text{if } b_2 = x, b_1 \in A & \text{or} \\ b_1 = b_2 = x. \end{cases}$$

Then B is easily seen to be well-ordered because R is a well-order, but $(A, R) < (B, \leq)$ in Y contradicting the maximality of (A, R). Thus $A = X$ and X can be well-ordered. □

Remark A.24. We have used (WO), hence an equivalent form of (AC), to prove Theorem 4.7. It turns out that this theorem implies (AC) so that Theorem 4.7 is also equivalent to (AC). This is established in the interesting and accessible article [15], in which the somewhat surprising fact that the general continuum hypothesis (GCH), stating that for any infinite cardinal $\mathfrak{m}$ no cardinal lies strictly between $\mathfrak{m}$ and $2^{\mathfrak{m}}$, implies (AC) is also proved.

A next step after this introductory text would be to study ordinals formally, in order to be able to use the transfinite version of induction. Cameron's text [7] is a nice source to get started with this.

Solutions to Exercises in the Text

Exercises from Chapter 1

1.1 (1) The set $\{a,b,c,b,a\}$ has 3 elements: a, b, and c. Repeating or permuting them in the list of elements does not change the set.

(2) The set $\{a,b,\{c,d\}\}$ also has 3 elements: a, b, and $\{c,d\}$, which is the set that has c and d as members. Note that c and d are not elements of $\{a,b,\{c,d\}\}$; only the set $\{c,d\}$ is.

(3) The set $\{a,b,\{a,b\}\}$ also has 3 elements: a, b, and $\{a,b\}$, which is the set that has a and b as members. Note that the set $\{a,b\}$ is a different object than repcating a and b as elements.

1.10 . (1). $\neg p \wedge q$ is true because $\neg p$ and q are both true. (2) $r \wedge \neg q$ is false because $\neg q$ is false. (3) $q \wedge \neg p$ is true as in (1).

1.14 We can, for instance, use a truth table:

p	q	$p \vee q$	$\neg p$	$\neg q$	$\neg p \wedge \neg q$	$(p \vee q) \vee (\neg p \wedge \neg q)$	$(p \vee q) \wedge (\neg p \wedge \neg q)$
T	T	T	F	F	F	T	F
T	F	T	F	T	F	T	F
F	T	T	T	F	F	T	F
F	F	F	T	T	T	T	F

1.22 The statement to negate reads as "the crime took place in the patio and either Colonel Mustard is guilty or the crime weapon was the knife." Its negation has the form

$$\begin{aligned}\neg(p \wedge (c \vee w)) &\equiv \neg p \vee \neg(c \vee w)\\ &\equiv \neg p \vee (\neg c \wedge \neg w),\end{aligned}$$

that is, "Either the crime did not take place in the patio or Mustard is innocent and the crime weapon is not the knife."

F. Mynard, *An Introduction to the Language of Mathematics*,
https://doi.org/10.1007/978-3-030-00641-9

1.23 (1)

p	q	r	$\neg r$	$q \wedge \neg r$	$p \vee (q \wedge \neg r)$
T	T	T	F	F	T
T	T	F	T	T	T
T	F	T	F	F	T
T	F	F	T	F	T
F	T	T	F	F	F
F	T	F	T	T	T
F	F	T	F	F	F
F	F	F	T	F	F

(2)

p	q	r	s	$\neg q \vee p$	$r \vee \neg s$	$(\neg q \vee p) \wedge (r \vee \neg s)$
T	T	T	T	T	T	T
T	T	T	F	T	T	T
T	T	F	T	T	F	F
T	T	F	F	T	T	T
T	F	T	T	T	T	T
T	F	T	F	T	T	T
T	F	F	T	T	F	F
T	F	F	F	T	T	T
F	T	T	T	F	T	F
F	T	T	F	F	T	F
F	T	F	T	F	F	F
F	T	F	F	F	T	F
F	F	T	T	T	T	T
F	F	T	F	T	T	T
F	F	F	T	T	F	F
F	F	F	F	T	T	T

1.26 (1): Let p be "f is differentiable at x" and q be "f is continuous at x." Then the statement is $p \Longrightarrow q$. (2): Its converse is $q \Longrightarrow p$, which is false. For instance, the function $f(x) = |x|$ is continuous at 0 (so q is true) but not differentiable at 0 (so p is false). (3): The contrapositive is $\neg q \Longrightarrow \neg p$, that is, "if f is not continuous at x, then f is not differentiable at x" which is equivalent to the original statement, hence true. (4): In view of (1.5), the negation has the form $p \wedge \neg q$, which reads as "f is differentiable at x and not continuous at x."

1.28 Let s denote "it snows tonight" and h denote "I will stay at home tomorrow." The statement has the form $s \Longrightarrow h$, so that its converse $h \Longrightarrow s$ reads "if I stay at home tomorrow, then it snows tonight" and its contrapositive $\neg h \Longrightarrow \neg s$ reads "if I do not stay home tomorrow, then it does not snow tonight."

1.29 We can build a truth table, or note that

$$
\begin{aligned}
((p\Longrightarrow q)\wedge(q\Longrightarrow r))\Longrightarrow(p\Longrightarrow r) &\equiv \neg((p\Longrightarrow q)\wedge(q\Longrightarrow r))\vee(p\Longrightarrow r)\\
&\equiv \neg(p\Longrightarrow q)\vee\neg(q\Longrightarrow r)\vee\neg p\vee r\\
&\equiv (p\wedge\neg q)\vee(q\wedge\neg r)\vee\neg p\vee r\\
&\equiv ((p\wedge\neg q)\vee\neg p)\vee((q\wedge\neg r)\vee r)\\
&\equiv ((p\vee\neg p)\wedge(\neg q\vee\neg p))\vee((q\vee r)\wedge(\neg r\vee r))\\
&\equiv (\mathrm{T}\wedge(\neg q\vee\neg p))\vee((q\vee r)\wedge\mathrm{T})\\
&\equiv \neg q\vee\neg p\vee q\vee r\\
&\equiv \mathrm{T}\vee(\neg p\vee r)\equiv\mathrm{T}.
\end{aligned}
$$

1.31 $P\equiv Q$ if they are identical functions of their arguments, that is, if they take the same truth values for the same arguments, in other words, if $P\iff Q$ is always true, that is, if $P\iff Q$ is a tautology.

1.33 Let c be "X is compact" and b be "X is bounded." Then "X is compact is sufficient for X to be bounded" is

$$c\Longrightarrow b.$$

Let c denote "G is a cyclic group" and a denote "G is an Abelian group." Then "A necessary condition for a group G to be cyclic is that G be Abelian" is

$$c\Longrightarrow a.$$

1.34 Let P denote "p is prime," Q denote "p divides $a\cdot b$," A denote "p divides a" and B denote "p divides b." Then the statement has the form

$$(P\wedge Q)\Longrightarrow(A\vee B).$$

1.52 (1) is true for there are real numbers, e.g., 2, that are not more than 3. (2) is true: for every real number x, the number $x+1$ is larger than x. (3) is false for $x=0$ is a real number that does not satisfy $|x|>0$. (4) Since the truth set in $\mathbb{R}$ of $2x-4=0$ is $\{2\}$ and 2 is less than 3, this is a true statement. (5) is false: in the reals, $x^2+9\geq 9>0$ so that there is no real number satisfying $x^2+9=0$. (6) is false for there are real numbers, say 3, that do not satisfy $(x<3\wedge 2x-4=0)$.

1.56 (1). Let $D(x)$ denote "x is developed" and $U(x)$ denote "x is underdeveloped" and $H(x,y)$ denote "x helps y." Then the statement has the form

$$\exists x\, D(x)\wedge(\forall y\, U(y)\Longrightarrow\neg H(x,y)).$$

(2). In view of (1.15) and Proposition 1.15,

$$\neg(\exists x\, D(x)\wedge(\forall y\, U(y)\Longrightarrow\neg H(x,y)))\equiv\forall x\,\neg D(x)\vee\neg(\forall y\, U(y)\Longrightarrow\neg H(x,y)),$$

and

$$\neg(\forall y\, U(y) \Longrightarrow \neg H(x,y)) \equiv \exists y\, U(y) \wedge H(x,y)$$

by (1.14) and (1.5). Thus the negation has the form

$$\forall x\, \neg D(x) \vee \exists y\, (U(y) \wedge H(x,y)),$$

which reads as "either a country is not developed or there is an underdeveloped country that it helps."

1.67

$$\mathbb{P}(\{1\}) = \{\emptyset, \{1\}\};\ \mathbb{P}(\{\alpha,\beta\}) = \{\emptyset, \{\alpha\}, \{\beta\}, \{\alpha,\beta\}\},$$

$$\mathbb{P}(\{1,2,3\}) = \{\emptyset, \{1\}, \{2\}, \{3\}, \{1,2\}, \{1,3\}, \{2,3\}, \{1,2,3\}\},$$

and

$$\begin{aligned}\mathbb{P}(\{a,b,c,d\}) = \Big\{&\emptyset, \{a\}, \{b\}, \{c\}, \{d\}, \{a,b\}, \{a,c\}, \{a,d\}, \{b,c\},\\ &\{b,d\}, \{c,d\}, \{a,b,c\}, \{a,c,d\}, \{a,b,d\}, \{b,c,d\}, \{a,b,c,d\}\Big\}.\end{aligned}$$

Note that $\mathbb{P}(\{1\})$ has 2 elements, $\mathbb{P}(\{\alpha,\beta\})$ has 4 elements, $\mathbb{P}(\{1,2,3\})$ has 8 elements, and $\mathbb{P}(\{a,b,c,d\})$ has 16 elements.

1.68 (1). The subsets of X with 2 elements are

$$\{1,2\}, \{1,3\}, \{1,4\}, \{2,3\}, \{2,4\}, \{3,4\}.$$

Hence

$$\mathscr{A} = \{\{1,2\}, \{1,3\}, \{1,4\}, \{2,3\}, \{2,4\}, \{3,4\}\}.$$

(2). (a) is False, for elements of $\mathscr{A}$ are subsets of X with 2 elements and 1 is not such a set. (b) is True because $\{1,2\}$ is such a set. (c) is False, for elements of $\{1,2\}$ (that is, 1 and 2) are not such sets, but (d) is True, for elements of $\{\{1,2\},\{2,3\}\}$ are $\{1,2\}$ and $\{2,3\}$, which are subsets of X with 2 elements, hence are elements of $\mathscr{A}$. (f) is False for $\emptyset$ is not a subset of X with 2 elements, but (g) is True, for $\emptyset$ is a subset of any set, in particular of the set $\mathscr{A}$.

1.69 If $x \in A$ then $x \in B$ because $A \subset B$, but then $x \in C$ because $B \subset C$. Hence $A \subset C$.

1.70 Assume $A \subset B$ and let $C \in \mathbb{P}(A)$, that is, $C \subset A$. By Exercise 1.69, $C \subset B$, that is, $C \in \mathbb{P}(B)$. Thus $\mathbb{P}(A) \subset \mathbb{P}(B)$. Conversely, if $\mathbb{P}(A) \subset \mathbb{P}(B)$, then $A \in \mathbb{P}(B)$ because $A \in \mathbb{P}(A)$. Thus $A \subset B$.

1.85 There are many such sequences. Take, for instance, $A_n = \{k \in \mathbb{N} : k \geq n\}$. Then $A_1 = \mathbb{N}$ so that $\bigcup_{n=1}^{\infty} A_n = \mathbb{N}$ and $\bigcap_{n=1}^{\infty} A_n = \emptyset$. Indeed, there is no $k \in \mathbb{N}$ with $k \geq n$ for every $n \in \mathbb{N}$.

1.88 $x \in (\bigcap_{A\in\mathscr{A}} A)^c$ if and only if $x \notin \bigcap_{A\in\mathscr{A}} A$, that is, if and only if there is $A \in \mathscr{A}$ with $x \notin A$, that is, if and only if $x \in \bigcup_{A\in\mathscr{A}} A^c$. Similarly $x \in (\bigcup_{A\in\mathscr{A}} A)^c$ if and only if $x \notin \bigcup_{A\in\mathscr{A}} A$ if and only if $x \in A^c$ for every $A \in \mathscr{A}$, that is, if $x \in \bigcap_{A\in\mathscr{A}} A^c$.

1.89 If $x \in A \cap \bigcup_{B\in\mathscr{B}} B$, then $x \in A$ and there is $B \in \mathscr{B}$ with $x \in B$, so that $x \in \bigcup_{B\in\mathscr{B}} A \cap B$. The reverse inclusion is clear.

If $x \in A \cup \bigcap_{B\in\mathscr{B}} B$, then $x \in A$ or for every $B \in \mathscr{B}$, $x \in B$. If $x \in A \subset A \cup B$, then $x \in \bigcap_{B\in\mathscr{B}}(A \cup B)$. On the other hand, $\bigcap_{B\in\mathscr{B}} B \subset \bigcap_{B\in\mathscr{B}}(A \cup B)$. Thus $A \cup \bigcap_{B\in\mathscr{B}} B \subset \bigcap_{B\in\mathscr{B}}(A \cup B)$. Finally, if for every $B \in \mathscr{B}$, $x \in A \cup B$, then either $x \in A \subset A \cup \bigcap_{B\in\mathscr{B}} B$ and we are done, or $x \notin A$ and then $x \in \bigcap_{B\in\mathscr{B}} B$.

1.133 (1)

$$\begin{aligned}
(x,y) \in A \times (B \cup C) &\iff x \in A \wedge y \in B \cup C \\
&\iff (x \in A) \wedge (y \in B \vee y \in C) \\
&\overset{(1.2)}{\iff} (x \in A \wedge y \in B) \vee (x \in A \wedge y \in C) \\
&\iff x \in (A \times B) \cup (A \times C).
\end{aligned}$$

(2)

$$\begin{aligned}
(x,y) \in A \times (B \cap C) &\iff x \in A \wedge y \in B \cap C \\
&\iff (x \in A) \wedge (y \in B \wedge y \in C) \\
&\iff (x \in A \wedge y \in B) \wedge (x \in A \wedge y \in C) \\
&\iff x \in (A \times B) \cap (A \times C).
\end{aligned}$$

(3) $(x,y) \in A \times \emptyset$ is false for every pair (x,y), for $y \in \emptyset$ is false for every y. Similarly $(x,y) \in \emptyset \times A$ is false. Hence $A \times \emptyset = \emptyset \times A = \emptyset$.

(4)

$$\begin{aligned}
(x,y) \in (A \times B) \cap (C \times D) &\iff (x,y) \in A \times B \wedge (x,y) \in C \times D \\
&\iff x \in A \cap C \wedge y \in B \cap D \\
&\iff (x,y) \in (A \cap C) \times (B \cap D).
\end{aligned}$$

(5)

$$\begin{aligned}
(x,y) \in (A \times B) \cup (C \times D) &\iff (x,y) \in A \times B \vee (x,y) \in C \times D \\
&\iff (x \in A \wedge y \in B) \vee (x \in C \wedge y \in D) \\
&\iff (x \in A \vee x \in C) \wedge (y \in B \vee y \in D). \\
&\iff (x,y) \in (A \cup C) \times (B \cup D).
\end{aligned}$$

1.138 $f[\{a,b,c\}] = \{1,5\}$,

$$f^{-1}[\{1,2\}] = f^{-1}[\{1\}] = \{a,b\} \text{ and } f^{-1}[\{2\}] = \emptyset.$$

Moreover,

$$f[\{a,c\}] \cap f[\{b,d\}] = \{1,5\} \cap \{1,6\} = \{1\}$$

but

$$f[\{a,c\} \cap \{b,d\}] = f[\emptyset] = \emptyset,$$

so that

$$f[\{a,c\} \cap \{b,d\}] \neq f[\{a,c\}] \cap f[\{b,d\}].$$

Finally, $f[A] = \{1,5,6\}$ and $f^{-1}[f[A]] = A$, while $f^{-1}[B] = A$ and $f[f^{-1}[B]] = f[A] = \{1,5,6\}$.

Exercises from Chapter 2

2.4 If x is odd, then there is $k \in \mathbb{Z}$ with $x = 2k-1$. Hence, $x+1 = 2k$ is even.

2.7 Suppose $n|a$ and $n|b$, that is, $a = nk$ and $b = np$, for some integers k and p. Then

$$a - b = nk - np = n(k-p)$$

and $k - p \in \mathbb{Z}$. Thus $n|a-b$.

2.15 (1). If $B \setminus A = B$ then if $x \in B$, then $x \in B = B \setminus A$ so that $x \notin A$. Hence $B \cap A = \emptyset$. Conversely, if $B \cap A = \emptyset$ and $x \in B$, then $x \notin A$ so that $x \in B \setminus A$. Hence $B \subset B \setminus A$ and the reverse inclusion is always true, so that $B = B \setminus A$.

(2). Assume $B \setminus (B \setminus A) = A$. If $x \in A$, then $x \in B \setminus (B \setminus A) \subset B$ so that $x \in B$. Hence $A \subset B$. Conversely, if $A \subset B$ and $x \in A$ then $x \in B$ and $x \in A$ so $x \notin B \setminus A$, that is, $x \in B \setminus (B \setminus A)$. In other words, $A \subset B \setminus (B \setminus A)$. If, on the other hand, $x \in B$ and $x \notin B \setminus A$, then $x \in A$ because $A \subset B$. Thus $B \setminus (B \setminus A) \subset A$.

(3) By the distributive laws,

$$\begin{aligned}(A \cup B) \cap (A \cup B^c) &= ((A \cup (B \cap B^c)) \\ &= A \cup \emptyset = A.\end{aligned}$$

(4). $A \subset B \Longrightarrow A \cap B = A$: In general $A \cap B \subset A$. If $A \subset B$, then $A \subset A \cap B$ and thus $A \cap B = A$.

$A \cap B = A \Longrightarrow A \cup B = B$: In general $B \subset A \cup B$. If $x \in A \cup B$ then $x \in A$ or $x \in B$. If $A \cap B = A$ then if $x \in A$, then $x \in B$. Either way $x \in B$. Thus $A \cup B = B$.

$A \cup B = B \Longrightarrow A \subset B$: If $x \in A$, then $x \in A \cup B = B$ so $x \in B$, that is, $A \subset B$.

2.13 (1) if g and f are both one-to-one and $g \circ f(x) = g \circ f(y)$, then $f(x) = f(y)$ because g is one-to-one, and thus $x = y$ because f is one-to-one.

(2) Since $f[X] = Y$ and $g[Y] = Z$, we conclude that $g \circ f(X) = g[f[X]] = g[Y] = Z$, that is, $g \circ f$ is onto.

(3) in view of (1) and (2), if f and g are both one-to-one and onto, so is $g \circ f$.

(4) If $g \circ f$ is onto, then $g[f[X]] = Z$. Since $f[X] \subset Y$, that means that $g[Y] = Z$, that is, g is onto.

(5) If $g \circ f$ is one-to-one and $f(x) = f(y)$, then $g(f(x)) = g(f(y))$ so that $x = y$ because $g \circ f$ is one-to-one. Thus f is one-to-one.

2.26 Suppose $7n+9$ is even, that is, $7n+9 = 2k$ for some integer k. Then, subtracting $6n+9$ on both sides yields

$$n = 2k - 6n - 9 = 2(k - 3n - 4) - 1 = 2p - 1,$$

where $p = k - 3n - 4$ is an integer. Hence n is odd.

2.28 We proceed by contrapositive. Assume that m is not even, that is, m is odd. Then there is an integer k with $m = 2k - 1$ and thus

$$m^2 = (2k-1)^2 = 4k^2 - 4k + 1$$

so that $m^2 = 2p + 1$ for the integer $p = 2k^2 - 2k$. Hence, m^2 is odd, that is, not even.

2.31 We prove the contrapositive. Assume $3x \leq y$ is false, that is, $y < 3x$ and $2y > x$, that is, $y - 3x < 0$ and $2y - x > 0$ so that

$$(y - 3x)(2y - x) = 2y^2 + 3x^2 - 7xy < 0,$$

that is, $7xy \leq 3x^2 + 2y^2$ is false.

2.32 (6) We show the contrapositive of the direction $\Longrightarrow$: Assume that $f[X \setminus A] \not\subset Y \setminus f[A]$ for some $A \subset X$, that is, there is $y \in f[X \setminus A]$ with $y \in f[A]$. Thus there is $a \in A$ and $x \in X \setminus A$ with $f(a) = y = f(x)$. Since $x \notin A$ and $a \in A$, $x \neq a$ so that f is not one-to-one.

To show the converse (the direction $\Longleftarrow$) assume that $f[X \setminus A] \subset Y \setminus f[A]$ for every $A \subset X$ and let $x \neq t$ in X. Consider $A = \{x\}$. Then $t \notin A$ so that $f(t) \in f[X \setminus A] \subset Y \setminus f[A]$, that is, $f(t) \notin f[A] = f(\{x\})$. Hence $f(t) \neq f(x)$ and f is one-to-one.

(7) If f is onto and $y \in Y \setminus f[A]$ then, since $Y = f[X]$, there is $x \in X \setminus A$ with $f(x) = y$, that is, $Y \setminus f[A] \subset f[X \setminus A]$. Assume conversely that $Y \setminus f[A] \subset f[X \setminus A]$ for every $A \subset X$ and let $y \in Y$. Let $A = \emptyset$. Then

$$Y \setminus f[A] \subset f[X \setminus A] \iff Y \subset f[X],$$

that is, f is onto.

(8) Let $y \in f[f^{-1}[B]]$. Then there is $x \in f^{-1}[B]$ with $f(x) = y$ and $f(x) \in B$ because $x \in f^{-1}[B]$. Hence $y \in B$.

(9) Assume that $B \subset f[f^{-1}[B]]$ for every $B \subset Y$. Let $B = Y$, then $Y \subset f[f^{-1}[Y]] = f[X]$, that is, f is onto. Assume conversely that f is onto and let $b \in B \subset Y$. Since f is onto, there is $x \in X$ with $f(x) = b$, so that $x \in f^{-1}[B]$ and $b \in f[f^{-1}[B]]$.

(10) If $A \subset X$ and $a \in A$, then $f(a) \in f[A]$, that is, $a \in f^{-1}[f[A]]$.

(11) Assume that $f^{-1}[f[A]] \subset A$ for every $A \subset X$. Suppose $f(x) = f(t)$ and let $A = \{x\}$. Then

$$\{x,t\} \subset f^{-1}[f[A]] \subset A = \{x\},$$

that is, $x = t$. Assume conversely that f is one-to-one and let $A \subset X$, and $x \in f^{-1}[f[A]]$, that is, $f(x) \in f[A]$, that is, there is $a \in A$ with $f(a) = f(x)$. Since f is one-to-one $a = x \in A$.

2.52 If $-2 < x < 1$, then $x+2 > 0$ and $x-1 < 0$ so that $(x-1)(x+2) < 0$. On the other hand, $x-3 < 0$ and $x+3 > 0$ so that $(x-3)(x+3) < 0$. As a result, $\dfrac{(x-1)(x+2)}{(x-3)(x+3)} > 0$.

Similarly, if $x > 3$, all 4 factors in $\dfrac{(x-1)(x+2)}{(x-3)(x+3)}$ are positive, so that

$$\frac{(x-1)(x+2)}{(x-3)(x+3)} > 0.$$

2.54 Assume $xy = 0$ and $x \neq 0$. Then $\frac{1}{x}$ exists and thus

$$y = y \cdot 1 = y \cdot \left(x \cdot \frac{1}{x}\right) = (yx) \cdot \frac{1}{x} = 0 \cdot \frac{1}{x} = 0.$$

2.56 In view of (2.12), it is equivalent to show that if $x \in A$ and $x \in B$, then $x \in A \cap B$, which is the definition of $A \cap B$.

2.64 (1) Take $x = 1$ and $y = -1$. Then $|x+y| = |0| = 0$ but $|x| + |y| = 2$.

(2) If $n = 1$, then $n^5 - n = 0$ is even, but n is odd.

(3) If $C = \emptyset$ and $A \neq B$, say, $A = \mathbb{R}$ and $B = \mathbb{N}$ then $A \times B = A \times C = \emptyset$ but $A \neq B$.

(4) Let $X = \{a,b\}$ and $Y = \{b,c\}$. Then $X \cup Y = \{a,b,c\}$ so that $|X \cup Y| = 3$ but $|X| + |Y| = 2 + 2 = 4$.

2.70 If you proceed from scratch (as opposed to getting a quick answer using (2.20)), you can observe that for

$$n = 1 \Longrightarrow \sum_{i=1}^{n} 2i = 2 = 1 + 1^2$$

$$n = 2 \Longrightarrow \sum_{i=1}^{n} 2i = 2 + 4 = 2 + 2^2$$

$$n = 3 \Longrightarrow \sum_{i=1}^{n} 2i = 2 + 4 + 6 = 12 = 3 + 3^2$$

$$n = 4 \Longrightarrow \sum_{i=1}^{n} 2i = 2 + 4 + 6 + 8 = 20 = 4 + 4^2$$

$$n = 5 \Longrightarrow \sum_{i=1}^{n} 2i = 30 = 5 + 5^2$$

and conjecture that

$$\sum_{i=1}^{n} 2i = n^2 + n \tag{A.7}$$

for all $n \in \mathbb{N}$. You would have then to prove it by induction:

We have already verified (A.7) for $n = 1$. Assume that (A.7) is true for some n. We want to show

$$\sum_{i=1}^{n+1} 2i = (n+1)^2 + (n+1).$$

To this end, note that

$$\begin{aligned}\sum_{i=1}^{n+1} 2i &= \sum_{i=1}^{n} 2i + 2(n+1) \overset{(A.7)}{=} n^2 + n + 2n + 2 \\ &= (n^2 + 2n + 1) + (n+1) = (n+1)^2 + (n+1).\end{aligned}$$

We conclude by induction that (A.7) is true for all $n \in \mathbb{N}$.

2.73 We proceed by induction on $n \geq 0$. If $n = 0$ then $n^3 - n = 0$ so that $3|n^3 - n$. Assume that $3|n^3 - n$ for some $n \geq 0$. Then

$$(n+1)^3 - (n+1) = n^3 + 3n^2 + 3n + 1 - n - 1 = (n^3 - n) + 3(n^2 + n).$$

Since $3|n^3 - n$ by inductive hypothesis and 3 divides $3(n^2 + n)$, we conclude that $3|(n+1)^3 - (n+1)$. By induction, $3|(n^3 - n)$ for all integers $n \geq 0$.

2.75 We proceed by induction. For $n = 1$, we check that $n + 3 = 4 < 5(1)^2$. Assume now that $n + 3 < 5n^2$ for some n. Then

$$5(n+1)^2 = 5(n^2 + 2n + 1) = 5n^2 + 10n + 5$$

so that using the inductive hypothesis that $5n^2 > n + 3$, we have

$$5(n+1)^2 > n + 3 + 10n + 5 = 11n + 8 > (n+1) + 4,$$

which completes the proof by induction.

2.76 We proceed by induction. For $n = 1$, $2^n = 2$ and $2^{n+1} - 2^{n-1} - 1 = 4 - 1 - 1 = 2$ so that the inequality is satisfied.

Assume that

$$2^n \leq 2^{n+1} - 2^{n-1} - 1 \tag{A.8}$$

is satisfied for some $n \in \mathbb{N}$. Multiplying both sides by 2 yields

$$2^{n+1} \leq 2^{n+2} - 2^n - 2$$

and since $a - 2 \leq a - 1$ for every a, in particular for $a = 2^{n+2} - 2^n$, we conclude that

$$2^{n+1} \leq 2^{n+2} - 2^n - 1.$$

By induction (A.8) is true for all $n \in \mathbb{N}$.

2.78 We proceed by induction. For $n = 1$, $\sum_{i=1}^{n} 2^i = 2^1 = 2$ and $2^{n+1} - 2 = 2^2 - 2 = 2$, so that the formula is verified. Assume now that

$$\sum_{i=1}^{n} 2^i = 2^{n+1} - 2 \tag{A.9}$$

for some natural number n. Then, since

$$\sum_{i=1}^{n+1} 2^i = \sum_{i=1}^{n} 2^i + 2^{n+1},$$

the inductive assumption (A.9) applies to the effect that

$$\sum_{i=1}^{n+1} 2^i = 2^{n+1} - 2 + 2^{n+1} = 2 \cdot 2^{n+1} - 2 = 2^{n+2} - 2,$$

which completes the proof by induction.

2.79 For $n = 0$, $\sum_{i=0}^{0} i \cdot i! = 0$ and $(0+1)! - 1 = 1 - 1 = 0$ so that the formula is verified. Assume now that

$$\sum_{i=0}^{n} i \cdot i! = (n+1)! - 1$$

is true for some integer $n \geq 0$. Then

$$\sum_{i=0}^{n+1} i \cdot i! = \sum_{i=0}^{n} i \cdot i! + (n+1) \cdot (n+1)!.$$

Using the inductive hypothesis, we obtain

$$\sum_{i=0}^{n+1} i \cdot i! = (n+1)! - 1 + (n+1) \cdot (n+1)!,$$

so that, factoring $(n+1)!$ in the first and last term,

$$\begin{aligned}\sum_{i=0}^{n+1} i \cdot i! &= (n+1)!\,(1 + (n+1)) - 1 \\ &= (n+1)!(n+2) - 1 \\ &= (n+2)! - 1,\end{aligned}$$

which completes the proof by induction.

2.84 We proceed by strong induction on $n \in \mathbb{N}$. For $n = 1$, this is $a - b = (a-b) \times 1$ and for $n = 2$, this is, $a^2 - b^2 = (a-b)(a+b)$. Assume that (2.28) is true for all k with $1 \leq k \leq n$ for a certain $n \in \mathbb{N}$.

Noting that

$$\begin{aligned} a^{n+1} - b^{n+1} &= a \cdot a^n - b \cdot b^n \\ &= (a^n - b^n)(a+b) + ab^n - a^n b \\ &= (a^n - b^n)(a+b) - ab(a^{n-1} - b^{n-1}), \end{aligned}$$

we can apply the inductive hypothesis to both $k = n$ and $k = n-1$ to the effect that

$$\begin{aligned} a^{n+1} - b^{n+1} &= (a+b)(a-b)(a^{n-1} + a^{n-2}b + \ldots + ab^{n-2} + b^{n-1}) \\ &\quad -ab(a-b)(a^{n-2} + a^{n-3}b + \ldots + ab^{n-3} + b^{n-2}) \\ &= (a-b)\Big((a^n + a^{n-1}b + \ldots + a^2b^{n-2} + ab^{n-1}) \\ &\quad +(ba^{n-1} + a^{n-2}b^2 + \ldots ab^{n-1} + b^n) \\ &\quad -ab(a^{n-2} + a^{n-3}b + \ldots + ab^{n-3} + b^{n-2})\Big) \\ &= (a-b)\left(a^n + a^{n-1}b + \ldots + a^2b^{n-2} + ab^{n-1} + b^n\right) \\ &\quad +(ba^{n-1} + a^{n-2}b^2 + \ldots ab^{n-1}) \\ &\quad -(a^{n-1}b + a^{n-2}b^2 + \ldots ab^{n-1}) \\ &= (a-b)\left(a^n + a^{n-1}b + \ldots + a^2b^{n-2} + ab^{n-1} + b^n\right). \end{aligned}$$

We conclude by induction that (2.28) is true for all $n \in \mathbb{N}$.

2.86 If $b < 0$ and $a < 0$, then we may divide $-a$ by $-b = |b|$ by the base case to the effect that there are integers q' and r' with $0 \leq r' < |b|$ such that

$$-a = -bq' + r'.$$

If $r' = 0$ then $a = bq'$ and we take $q = q'$ and $r = r' = 0$. Otherwise, $r' > 0$ and thus

$$a = bq' - r' = (q'+1)b - b - r' = (q'+1)b + |b| - r'$$

and thus $a = qb + r$ for $q = q' + 1$ and $r = |b| - r'$, which satisfies $0 \leq r < |b|$.

Similarly, if $a \geq 0$ and $b < 0$, we divide a by $-b = |b|$ to the effect that there are q' and r with

$$a = -bq' + r,\ 0 \leq r < |b|.$$

Letting $q = -q'$ yields the desired result.

If $a < 0$ and $b > 0$, we can divide $-a$ by b to the effect that there are q' and r' with

$$-a = bq' + r',\ 0 \leq r' < b.$$

If $r' = 0$ then $q = -q'$ and $r = r' = 0$ yields the desired result. Otherwise,

$$a = -bq' - r' = (-q' - 1)b + b - r' = qb + r,$$

where $q = (-q' - 1)$ and $r = b - r'$ satisfies $0 \leq r < b$.

2.81 For $n = 1$, the statement is true, but not for $n = 2$. In the inductive step part of the "proof," it is implicitly assumed that $n \geq 2$, so that $n+1 \geq 3$ and removing one ball leaves of group of at least 2 balls. The argument would work in this case. But if $n = 1$ and $n+1 = 2$, removing a ball leaves only one ball and the argument breaks down.

2.96 We proceed by induction. For $n = 1$, $\sum_{k=1}^{1} F_{2k-1} = F_1 = 1$ and $F_{2n} = F_2 = 1$. Similarly, for $n = 2$,

$$\sum_{k=1}^{n} F_{2k-1} = \sum_{k=1}^{2} F_{2k-1} = F_1 + F_3 = 1 + 2 = 3$$

and $F_{2n} = F_4 = 3$.

Assume that $\sum_{k=1}^{n} F_{2k-1} = F_{2n}$ for some natural number n. Now

$$\begin{aligned}\sum_{k=1}^{n+1} F_{2k-1} &= \sum_{k=1}^{n} F_{2k-1} + F_{2n+1} \\ &= F_{2n} + F_{2n+1}\end{aligned}$$

by inductive hypothesis. In view of (2.32),

$$\sum_{k=1}^{n+1} F_{2k-1} = F_{2n+2}.$$

We conclude by induction that $\sum_{k=1}^{n} F_{2k-1} = F_{2n}$ for every $n \in \mathbb{N}$.

2.97 We proceed by induction. If $n = 1$, then $\sum_{k=1}^{n} F_k^2 = F_1^2 = 1$ and $F_n \cdot F_{n+1} = F_1 F_2 = 1$. For $n = 2$,

$$\sum_{k=1}^{n} F_k^2 = F_1^2 + F_2^2 = 1 + 1 = 2$$

and $F_n \cdot F_{n+1} = F_2 F_3 = 2$.

Assume now that (2.34) for some $n \in \mathbb{N}$. Then

$$\begin{aligned}\sum_{k=1}^{n+1} F_k^2 &= \sum_{k=1}^{n} F_k^2 + (F_{n+1})^2 \\ &= F_n F_{n+1} + F_{n+1}^2\end{aligned}$$

by inductive hypothesis. Factoring F_{n+1} and using (2.32) yields

$$\sum_{k=1}^{n+1} F_k^2 = F_{n+1} \cdot (F_n + F_{n+1}) = F_{n+1} \cdot F_{n+2}.$$

We conclude by induction that (2.34) is true for all $n \in \mathbb{N}$.

2.99 We proceed by induction. This is true for $n=1$ and $n=2$ because $F_5=5$ and $F_{10}=55$. Assume now that $5|F_{5n}$ for some $n\in\mathbb{N}$. Then

$$\begin{aligned} F_{5(n+1)} &= F_{5n+5} = F_{5n+4}+F_{5n+3} = F_{5n+3}+F_{5n+2}+F_{5n+2}+F_{5n+1} \\ &= F_{5n+2}+F_{5n+1}+2(F_{5n+1}+F_{5n})+F_{5n+1} \\ &= F_{5n+1}+F_{5n}+4F_{5n+1}+2F_{5n} \\ &= 5F_{5n+1}+2F_{5n}, \end{aligned}$$

so that $5|F_{5(n+1)}$ for $5|(5F_{5n+1})$ and $5|2F_{5n}$ because $5|F_{5n}$.

We conclude by induction that $5|F_{5n}$ for all $n\in\mathbb{N}$.

Exercises from Chapter 3

3.6 Assume that $R\subset S$ and let $(x,t)\in R^{-1}$, equivalently, $(t,x)\in R\subset S$, so that $(t,x)\in S$, equivalently, $(x,t)\in S^{-1}$. Hence $R^{-1}\subset S^{-1}$. Thus if $R^{-1}\subset S^{-1}$ then $(R^{-1})^{-1}\subset(S^{-1})^{-1}$, that is, $R\subset S$, showing the converse ([4]).

3.16 (1) R is not reflexive because $(c,c)\notin R$. It is not symmetric because $(a,c)\in R$, that is, aRc but $c \not R\, a$ because $(c,a)\notin R$. It is not antisymmetric either. Indeed, aRb and bRa but $a\neq b$. Finally, R is not transitive because cRb and bRa but $c \not R\, a$.

(2) R is reflexive (a number has the same digits as itself), symmetric (two numbers that have a digit in common do regardless of the order considered for these two numbers), but neither transitive (for instance, $1R19$ and $19R99$ but $1 \not R\, 99$) nor antisymmetric ($1R19$ and $19R1$ but $1\neq 19$).

(3) R is not reflexive. In fact $x<x$ is never true. R is not symmetric for $1<2$ but $2\not<1$, but it is antisymmetric because

$$x<y \text{ and } y<x$$

is always false, so that the premises of the conditional statement (3.2) are false. As a result (3.2) is a true statement. Moreover, R is transitive: $x<y$ and $y<z$ implies that $x<z$. It is neither an order nor an equivalence.

(4) $\leq$ is an order (reflexive, antisymmetric, transitive) just as in Example 3.15.

(5) The inclusion relation $\subset$ is a relation on $\mathbb{P}X$: ARB if $A\subset B$, that is, if every element of A is an element of B. This relation is reflexive ($A\subset A$ for every set A) and transitive (if $A\subset B$ and $B\subset C$, then $A\subset C$ as observed in Exercise 1.69). It is not symmetric. For instance, if $X=\{1,2,3\}$, $A=\{1\}$ and $B=\{1,2\}$, then $A\subset B$

[4] Note that we only really needed to show

$$R\subset S \Longrightarrow R^{-1}\subset S^{-1}$$

for general relation R and S, for applying this statement replacing R by R^{-1} and S by S^{-1} yields the converse.

but $B \not\subset A$. It is antisymmetric: if $A \subset B$ and $B \subset A$, then $A = B$, as noted in (1.18). Hence $\subset$ is reflexive, transitive, and antisymmetric, hence an order on $\mathbb{P}X$.

(6) The relation $=$ is the identity relation on $\mathbb{Z}$. As observed before, it is reflexive, symmetric, antisymmetric, and transitive, hence it is both an order and an equivalence relation.

(7) The relation R is reflexive (one has the same height as oneself), symmetric (that two people have the same height does not depend on the order in which we consider these two people), and transitive (if x and y have the same height and y and z have the same height, then x and z have the same height). Hence R is an equivalence relation. It is not antisymmetric: two different people x and y with the same height satisfy xRy and yRx but $x \neq y$.

(8) R is reflexive, for a set has the same number of elements as itself, symmetric for having the same number of elements does not depend on the order in which we consider the two sets, and transitive because if A and B have the same number of elements, and so do B and C, then A and C have the same number of elements. Hence R is an equivalence relation. It is not antisymmetric because it is not the identity and it is reflexive and symmetric (see Corollary 3.13).

(9) Every integer divides itself, so that $|$ is reflexive. It is not symmetric for $2|6$ but $6 \nmid 2$. It is transitive by Proposition 2.5. It is not antisymmetric on $\mathbb{Z}$: for instance, $2|-2$ and $-2|2$ but $2 \neq -2$.

(10) The relation "divides" is reflexive and transitive, and fails to be symmetric as in (9). On the other hand, it is antisymmetric on $\mathbb{N}$. Indeed, if $a|b$ and $b|a$, then there are n and p in $\mathbb{N}$ with $b = na$ and $a = pb$ so that $a = (pn)a$. As a result, $pn = 1$, so that $p = n = 1$ because p and n are natural numbers. Thus $a = b$. Hence $|$ is an order relation on $\mathbb{N}$ (but not on $\mathbb{Z}$).

3.17 (1) If R is reflexive, then $(x,x) \in R$, equivalently, $(x,x) \in R^{-1}$ for every $x \in X$, that is, R^{-1} is reflexive too.

(2) Assume R is transitive and let $xR^{-1}y$ and $yR^{-1}z$, equivalently, zRy and yRx. By transitivity of R, we conclude that zRx, that is, $xR^{-1}z$. Thus R^{-1} is transitive.

(3) If R is symmetric, $R = R^{-1}$, hence R^{-1} is symmetric.

(4) Assume R is antisymmetric and let $x, y \in X$ be such that $xR^{-1}y$ and $yR^{-1}x$, equivalently yRx and xRy. Since R is antisymmetric, $y = x$. Hence R^{-1} is antisymmetric.

3.24 (1) If R is functional, then $R(x)$ is a singleton for each $x \in X$, so that in particular $R(x) \neq \emptyset$ for every $x \in X$, that is, R^{-1} is surjective. On the other hand, R^{-1} is injective for if $x \in R^{-1}(y_1) \cap R^{-1}(y_2)$ then $\{y_1, y_2\} \subset R(x)$ so that $y_1 = y_2$ for $R(x)$ is a singleton.

Conversely, if R^{-1} is surjective and injective, then for every $x \in X$, $R(x) \neq \emptyset$ because R^{-1} is surjective, and $R(x)$ is a singleton because R^{-1} is injective: if $y_1, y_2 \in R(x)$, then $x \in R^{-1}(y_1) \cap R^{-1}(y_2)$ and thus $y_1 = y_2$.

(2) Suppose R is a functional relation, that is, $R(x) = \{r(x)\}$ is a singleton for all $x \in X$. The relation R is surjective if and only if for every $y \in Y$, there is $x \in X$ with $y \in R(x)$, equivalently, $y = r(x)$. In other words, R is surjective if and only if $r: X \to Y$ is a surjective function.

On the other hand, R is injective if and only if $x_1 \neq x_2 \Longrightarrow R(x_1) \cap R(x_2) = \emptyset$, that is, if and only if

$$x_1 \neq x_2 \Longrightarrow r(x_1) \neq r(x_2),$$

that is, if and only if $r : X \to Y$ is an injective function.

(3) Let $f : X \to Y$ be a one-to-one function. Then $\tilde{f}^{-1} : f[X] \to X$ is functional: for every $y \in f[X]$, there is $x \in X$ with $y = f(x)$ and x is unique because f is one-to-one. Thus $\tilde{f}^{-1}(y) = \{x\}$ is a singleton and thus $\tilde{f}^{-1}$ is functional.

3.26 As $z(S \circ R)x$ if and only if there is $y \in Y$ with zSy and yRx, this is equivalent to $y = r(x)$ and $z = s(y)$, that is, $z = s \circ r(x)$.

3.27 (1) By definition $z \in (S \circ R)(A)$ if there is $a \in A$ with $a(S \circ R)z$, that is, if there is $y \in Y$ with aRy and ySz, that is, if there is $y \in R(A)$ with $z \in R(y)$, equivalently, if $z \in S(R(A))$.

(2) By definition $(z,x) \in (S \circ R)^{-1}$ if and only if $(x,z) \in S \circ R$ if and only if there is $y \in Y$ with $(x,y) \in R$ and $(y,z) \in S$, equivalently, $(y,x) \in R^{-1}$ and $(z,y) \in S^{-1}$, that is, $(z,x) \in R^{-1} \circ S^{-1}$.

(3) $(x,t) \in R \circ I_X$ if and only if there is $s \in X$ with $(x,s) \in I_X$ and $(s,t) \in R$. But if $(x,s) \in I_X$ then $x = s$. Thus $(x,t) \in R$. A similar argument applies to $I_Y \circ R$.

3.28 Assume that R is transitive and let $(x,z) \in R \circ R$. Then there is $y \in X$ with xRy and yRz. Thus xRz by transitivity of R. Thus $(x,z) \in R$ and $R \circ R \subset R$. Assume conversely that $R \circ R \subset R$ and let x, y, and z be such that xRy and yRz. Then $(x,z) \in R \circ R \subset R$ so that $(x,z) \in R$, that is, xRz. Thus R is transitive.

3.46 The relation $\preceq$ is reflexive for a word is a prefix of itself (using the empty word for S in the definition of prefix). It is transitive for if $W \preceq Y$ and $Y \preceq Z$, then we have 4 cases to examine:

1. W is a prefix of Y and Y is a prefix of Z, in which case $Y = W \frown S$ and

$$Z = Y \frown U = W \frown S \frown U$$

   for some words S and U. As a result, W is a prefix of Z and thus $W \preceq Z$.
2. W is a prefix of Y, that is, $Y = W \frown S$ for some word S, and there are letters $a < b$ in A and words U, V, and R with $Y = U \frown a \frown V$ and $Z = U \frown b \frown R$. Hence W is either a prefix of U, hence of Z, or W contains $U \frown a$ as a prefix in which case W takes the form $U \frown a \frown V'$. Either way, $W \preceq Z$.
3. $W = U \frown a \frown V$ and $Y = U \frown b \frown R$ for some words U, V, and R and letters $a < b$ in A, and Y is a prefix of Z. Then

$$Z = Y \frown S = U \frown b \frown R \frown S$$

   for some word S, so that $W \preceq Z$.
4. $W = U \frown a \frown V$ and $Y = U \frown b \frown R$ for some words U, V and R and letters $a < b$ in A, and $Y = S \frown c \frown T$ and $Z = S \frown d \frown F$ for some words S, T, and F and letters $c < d$ in A. Because

$$Y = U \frown b \frown R = S \frown c \frown T$$

either $U = S$, $b = c$ and $R = T$ in which case $W \preceq Z$, or c is a letter in U, that is, $U = S \frown c \frown B$ for some word B, in which case

$$W = S \frown c \frown B \frown a \frown V = S \frown c \frown C$$

and $Z = S \frown d \frown F$ and $W \preceq Y$, or c is a letter in R, that is, $S = U \frown b \frown D$ so that

$$Z = U \frown b \frown D \frown d \frown F = U \frown b \frown E$$

and $W = U \frown a \frown V$ so that $W \preceq Z$.

Remains to see that $\preceq$ is antisymmetric. To this end, assume that $W \preceq Z$ and $Z \preceq W$. We again have several possibilities to examine:

1. If W is a prefix of Z and Z a prefix of W, then $Z = W$.
2. If $Z \preceq W$ because

$$Z = U \frown a \frown V \text{ and } W = U \frown b \frown R$$

for some words, U, V and R and letter $a < b$ in A, then W cannot be a prefix of Z for it starts with $U \frown b$. Thus

$$W = B \frown c \frown C \text{ and } Z = B \frown d \frown D \tag{A.10}$$

for some words B, C, and D and letters $c < d$ in A. If B is shorter than U, it is a prefix of U, that is, $U = B \frown Q$ for some word Q so that

$$W = B \frown Q \frown b \frown R \text{ and } Z = B \frown Q \frown a \frown V,$$

which is not compatible with (A.10). Indeed, Q would have to be empty, but then the letter after B in W would be $c = b < d = a$ which is not possible because $a < b$. The remaining cases being similar, only the first case applies and $Z = W$.

Finally, we show that if $\leq$ is a total order on A then $\preceq$ is a total order on X. If W is the empty word, then W is a prefix of any other words, hence $W \preceq Y$ for any word $Y \in X$. If W and Z are different non-empty words, they both have first letters w and z. Because $\leq$ is a total order, if $w \neq z$, then either $w < z$ in which case $W \preceq Z$ or $w > z$ in which case $W \succeq Z$. If $w = z$, we may apply the same argument to the first letter in which W and Z differ to the effect that W and Z are comparable under $\preceq$.

3.49 The relation $\preceq$ is reflexive because $x \leq_X x$ and $y \leq_Y y$. It is transitive because if $(x,y) \preceq (r,s)$ and $(r,s) \preceq (t,z)$ then $x \leq_X r$ and $r \leq_X t$ so that $x \leq_X t$ and $y \leq_Y s$ and $s \leq_Y z$ so that $y \leq_Y z$. Thus $(x,y) \preceq (t,z)$. It is antisymmetric for if $(x,y) \preceq (r,s)$ and $(r,s) \preceq (x,y)$ then $x \leq_X r$ and $r \leq_X x$ so that $x = r$, and $y \leq_Y s$ and $s \leq_Y y$ so that $s = y$, and $(x,y) = (r,s)$.

Similarly, $\preceq_\ell$ is reflexive for $y \leq_Y y$ for all $y \in Y$ so that $(x,y) \preceq_\ell (x,y)$ for all $(x,y) \in X \times Y$. It is transitive: assume $(x_1,y_1) \preceq_\ell (x_2,y_2)$ and $(x_2,y_2) \preceq_\ell (x_3,y_3)$ and examine the 4 cases:

1. $x_1 = x_2$ and $y_1 \leq_Y y_2$ and $x_2 = x_3$ and $y_2 \leq_Y y_3$. Then $x_1 = x_3$ and $y_1 \leq_Y y_3$ so that $(x_1,y_1) \preceq_\ell (x_3,y_3)$;
2. $x_1 = x_2$ and $y_1 \leq_Y y_2$ and $x_2 <_X x_3$, so that $x_1 <_X x_3$ and $(x_1,y_1) \preceq_\ell (x_3,y_3)$;
3. $x_1 <_X x_2$ and $x_2 = x_3$ and $y_2 \leq_Y y_3$, so that $x_1 <_X x_3$ and $(x_1,y_1) \preceq_\ell (x_3,y_3)$;
4. $x_1 <_X x_2$ and $x_2 <_X x_3$ so that $x_1 <_X x_3$ and $(x_1,y_1) \preceq_\ell (x_3,y_3)$.

Finally $\preceq_\ell$ is antisymmetric. If $(x_1,y_1) \preceq_\ell (x_2,y_2)$ and $(x_2,y_2) \preceq_\ell (x_1,y_1)$, then $x_1 <_X x_2$ is not possible for we could not have $(x_2,y_2) \preceq_\ell (x_1,y_1)$. Hence $x_1 = x_2$ and $y_1 \leq_Y y_2$ and $y_2 \leq_Y y_1$, so that $y_1 = y_2$.Thus $(x_1,y_1) = (x_2,y_2)$.

3.54 This is an order relation on $X = \{a,b,c,d,e,f,g\}$. Because it is an order, it is reflexive (loops at each vertex are not represented in the Hasse diagram) and transitive. Hence, for instance, aRg because aRd and dRg (only comparisons with immediate predecessors or successors are represented in the Hasse diagram). Thus the relation is

$$R = \Big\{(a,a),(b,b),(c,c),(d,d),(e,e),(f,f),(g,g),$$
$$(a,c),(a,d),(b,d),(b,e),(d,f),(d,g),(c,g),$$
$$(a,g),(a,f),(b,f)\Big\}.$$

3.56 (1) $\mathbb{N}$ does not have any upper bound, for there is no real number greater than every natural number. On the other hand, every real number of $(-\infty,1]$ is a lower bound. Hence, 1 is the greatest lower bound of $\mathbb{N}$, which is also its least element. Since it has no upper bound, it has no greatest element and no least upper bound.

(2) $\mathbb{Z}$ has no upper bound and no lower bound, hence no least nor greatest element, and no least upper bound or greatest lower bound.

(3) Every real number in $L = (-\infty,1]$ is a lower bound for the interval $(1,2)$ and every real number of $U = [2,\infty)$ is an upper bound for $(1,2)$. Hence, $(1,2)$ has greatest lower bound 1 and least upper bound 2. It has no greatest or least element.

(4) Every real number in $L = (-\infty,1]$ is a lower bound for the interval $[1,2)$ and every real number of $U = [2,\infty)$ is an upper bound for $[1,2)$. Hence, $[1,2)$ has greatest lower bound $1 \in [1,2)$, so that 1 is also the least element. On the other hand, the least upper bound is $2 \notin [1,2)$, so that $[1,2)$ has no greatest element.

(5) Every real number in $L = (-\infty,1]$ is a lower bound for the interval $(1,2]$ and every real number of $U = [2,\infty)$ is an upper bound for $(1,2]$. Hence, $(1,2]$ has greatest lower bound $1 \notin (1,2]$, so $(1,2]$ has no least element. On the other hand, the least upper bound is $2 \in (1,2]$, so that 2 is also the greatest element.

(6) L is the set of lower bounds, U that of upper bounds, the greatest lower bound is 1 and is the least element, the least upper bound is 2 and is the greatest element.

3.59 (1). Since 1, 2, and 3 divide 6, the number 6 is an upper bound for A and is in A, so that 6 is the greatest element of A. In contrast, 4 does not divide 6, and thus B does not have a greatest element. As 1 divides all elements of A and is in A, 1 is the least element of A. In contrast, B does not have a least element, for 4 does not divide 6.

(2). Because A has a greatest element, it has a unique maximal element, which is 6. Similarly, it has 1 as least element, which is therefore the unique minimal element. In contrast, 4 and 6 are both maximal and minimal in B, for there is no element of B other than 4 that divides 4 or is divided by 4, and similarly for 6.

(3). A lower bound for A is a natural number that divides all elements of A. There is only one such number: 1. An upper bound for A is a natural number that is divided by all elements of A. This is the case for all multiples of 6. On the other hand, the lower bounds for B are 1 and 2, for they are the only numbers that divide all the elements of B. The upper bounds for B are multiples of 12, for they are the numbers that both 4 and 6 divide.

(4) In view of (3), the greatest lower bound of A is 1 and of B is 2. The least upper bound of A is 6 and of B is 12.

3.61 (1). Since $\{a,b,c\}$ contains $\{a,b\}$ and $\{b,c\}$, the element $\{a,b,c\}$ is maximal in A, but the other elements are not. On the other hand, $\{a,b\}$ and $\{a,c\}$ are minimal, for there is no other element of A below them.

(2). $\{a,b,c\}$ is the greatest element of A for it is an upper bound for A that belongs to A. On the other hand, A has no least element, for there is no element of A that is contained in both $\{a,b\}$ and $\{a,c\}$.

(3). Upper bounds (in $\mathbb{P}X$) for A are $\{a,b,c\}$ and $\{a,b,c,d\}$ for these are the two elements of $\mathbb{P}X$ that are supersets of every element of A. Elements of $\mathbb{P}X$ that are subsets of every element of A are $\{a\}$ and $\emptyset$: they are the lower bounds of A.

(4). In view of (3), the greatest lower bound of A is $\{a\}$ (because $\emptyset \subset \{a\}$) and the least upper bound of A is $\{a,b,c\}$ (because $\{a,b,c\} \subset \{a,b,c,d\}$).

3.62 If $\mathscr{A} \subset \mathbb{P}X$ then

$$\sup_{A\in\mathscr{A}} A = \bigcup_{A\in\mathscr{A}} A \text{ and } \inf_{A\in\mathscr{A}} A = \bigcap_{A\in\mathscr{A}} A.$$

Indeed, $C \in \mathbb{P}X$ is an upper bound of $\mathscr{A}$ if and only if $A \subset C$ for every $A \in \mathscr{A}$, equivalently, if $\bigcup_{A\in\mathscr{A}} A \subset C$. Hence the least upper bound is $\bigcup_{A\in\mathscr{A}} A$. Similarly, $C \in \mathbb{P}X$ is a lower bound of $\mathscr{A}$ if and only if $C \subset A$ for every $A \in \mathscr{A}$, equivalently, if $C \subset \bigcap_{A\in\mathscr{A}} A$. Hence the greatest lower bound is $\bigcap_{A\in\mathscr{A}} A$.

3.64 (1). Since $\bigvee_{x\in X} x$ exists in X, it is necessarily the greatest element of X.

(2). If $A \subset X$, let L be the (possibly empty) set of lower bounds of A. Then $\bigvee L$ is by definition the greatest lower bound of A ([5]). (3) follows either from the footnote, or from the observation that from (2), $\bigwedge_{x\in X} x$ exists in X and is thus necessarily the least element.

3.67 (1) Since $f(x) \leq f(x)$ for all $x \in [a,b]$, the relation is reflexive on F. If $f \leq g$ and $g \leq h$, then for every $x \in [a,b]$, $f(x) \leq g(x) \leq h(x)$ so that $f(x) \leq h(x)$ because $\leq$ is transitive on $\mathbb{R}$. Hence $\leq$ is transitive on F. If $f \leq g$ and $g \leq f$, then for

[5] If L were empty, then $\bigvee L = \bigvee \emptyset$ would be the least element of X, hence an element of L. Indeed, any element of X is an upper bound of $\emptyset$, so that stating the existence of $\bigvee \emptyset$ is stating the existence of a least element of X. Thus L in fact is never empty.

every $x \in [a,b]$, $f(x) \leq g(x)$ and $g(x) \leq f(x)$, so that $f(x) = g(x)$ because $\leq$ is antisymmetric on $\mathbb{R}$. Hence $\leq$ is antisymmetric on F and is thus an order relation on F.

(2) This is not a total order. For instance, the function $f(x) = 1 - \frac{x-a}{b-a}$ and $g(x) = \frac{x-a}{b-a}$ are not comparable because $f(a) = 1 > g(a) = 0$ and $f(b) = 0 < g(b) = 1$.

(3) If $A \subset C$, then f_A and h_A defined on $[a,b]$, respectively, by $f_A(x) = \sup_{g\in A} g(x)$ and $h_A(x) = \inf_{g\in A} g(x)$ are well-defined functions because $[0,1]$ is a complete lattice. The functions f_A and h_A are easily seen to be the supremum and infimum of A in C, respectively.

(4) The greatest element of C is the constant function 1, while the least element is the constant function 0.

3.68 The set A does not have a greatest element for if $a \in A$ there is $n \in \mathbb{N}$ with $a = 1 - \frac{1}{n}$ but then $a' = 1 - \frac{1}{n+1} > a$ and $a' \in A$. On the other hand, 1 is the least upper bound of A. Indeed, 1 is an upper bound, and if $m < 1$, then there is $n \in \mathbb{N}$ with $\frac{1}{n} < 1 - m$ so that $m < 1 - \frac{1}{n}$, that is, m is not an upper bound for A.

The set B does not have a greatest element: for every $r \in B$, there is $r' \in B$ with $r' > r$. Indeed, if $r \in B$ and $r \leq 1$, then $r < \frac{6}{5}$ and $\frac{6}{5} \in B$ because $\frac{36}{25} < 2$. Assume that $1 < r \in \mathbb{Q}$ and $r^2 < 2$. Then $2 - r^2 > 0$. Let $m \in \mathbb{N}$ be large enough for $\frac{5}{m} < 2 - r^2$. Then

$$\left(r + \frac{1}{m}\right)^2 = r^2 + \frac{2r}{m} + \frac{1}{m^2} \leq r^2 + \frac{5}{m}$$

because, as $r > 1$, $r < r^2 < 2$ so that $\frac{2r}{m} \leq \frac{4}{m}$ and $\frac{1}{m^2} \leq \frac{1}{m}$. Since $r^2 + \frac{5}{m} < 2$, we conclude that $r' = r + \frac{1}{m} > r$ satisfies $r'^2 < 2$, that is, $r' \in B$.

The set B does not have a least upper bound. Indeed, if $m \in \mathbb{Q}$ is an upper bound for B, then $m > 1$ and $m^2 \geq 2$. Indeed, if $m^2 < 2$, then $m \in B$ and by the argument above, m is not an upper bound for B. If $m^2 > 2$, then by a similar argument as the one showing that B has no greatest element, there is another rational $m' > 1$ with $m^2 > m'^2 > 2$ so that m is not the least upper bound. Thus, if m were the least upper bound for B, then $m^2 = 2$. But there is no solution to this equation in $\mathbb{Q}$.

Remark A.25. Note that B has upper bounds, but no least upper bound. We will see in Section 3.4 that this is not possible in $\mathbb{R}$, which is an essential property of the reals to develop Analysis.

3.79 The relation R is clearly reflexive (one has the same age as oneself), symmetric (that x and y have the same age is symmetric in x and y), and transitive, for x and y have the same age and y and z have the same age, then x and y have the same age. Hence R is an equivalence relation.

3.80 R is an equivalence relation: it is reflexive (a line is parallel to itself), symmetric, and transitive (if x and y are parallel and y and z are parallel, then x and z are parallel). In contrast, S is not reflexive, is symmetric, and is not transitive (for instance, the lines L_1 and L_2 of equations $x = 0$ and $y = 0$ are S-related, and L_2 is S-related to the line L_3 of equation $x = 2$, but L_1 and L_3 are not S-related).

3.84 Since $\mathscr{P}$ is a partition, each $x \in X$ belongs to exactly one $P_x \in \mathscr{P}$. Of course, $x \sim_{\mathscr{P}} x$ because $x \in P$. Hence $\sim_{\mathscr{P}}$ is reflexive. By definition, $\sim_{\mathscr{P}}$ is symmetric. Moreover, $\sim_{\mathscr{P}}$ is transitive for if $x \sim_{\mathscr{P}} y$ and $y \sim_{\mathscr{P}} z$ then $P_x = P_y$ and $P_y = P_z$ and thus $P_x = P_z$ so that $x \sim_{\mathscr{P}} z$. By definition, the equivalence class of x for $\sim_{\mathscr{P}}$ is the unique $P_x \in \mathscr{P}$ for which $x \in P_x$.

3.90 $\equiv_p$ is reflexive, for p divides 0. It is symmetric, for if $p|(x-y)$ then $p|(y-x)$. It is transitive, for if $p|(x-y)$ and $p|(y-z)$ then p divides the sum

$$x-y+y-z=x-z,$$

that is, $x \equiv_p z$. Hence $\equiv_p$ is an equivalence relation. Note that $y \equiv_p x$ if and only if x and y have the same remainder in their Euclidean division by p. Indeed, if $x = pq_x + r$ and $y = pq_y + r$, then $x - y = p(q_x - q_y)$ so that $x \equiv_p y$. Conversely, if $x \equiv_p y$ and we consider divisions of x and y by p to the effect that $x = pq_x + r_x$ and $y = pq_y + r_y$ then

$$x-y=(q_x-q_y)p+(r_x-r_y)$$

is a multiple of p because $x \equiv_p y$. Hence

$$r_x - r_y = x - y + (q_y - q_x)p$$

is also a multiple of p. But $0 \le r_x, r_y < p$ so that $r_x - r_y$ cannot be a multiple of p, unless $r_x - r_y = 0$, that is, $r_x = r_y$.

As a consequence, the equivalence class of a given number x is the set of integers with the same remainder in the division by p, namely

$$[x] = \{x + np : n \in \mathbb{Z}\}.$$

On the other hand, the quotient set can be identified with the possible remainders in the division by p, that is,

$$\mathbb{Z}/\equiv_p \approx \{0,1,2,\dots p-1\},$$

where the canonical surjection q associates to $x \in \mathbb{Z}$ the remainder in the division of x by p.

3.108 Let C and D be cuts, and let

$$C+D = \{c+d : c \in C, d \in D\}.$$

If $x \in C+D$ and $y \le x$ then $y \le c+d = x$ for some $c \in C$ and some $d \in D$. Then $y - c \le d$ so that $y - c \in D$ because D is a cut. Hence, there is $d' \in D$ with $y - c = d'$, hence $y = c + d' \in C+D$.

Moreover, $C+D$ has no greatest element. Indeed, if $x \in C+D$, then there are $c \in C$ and $d \in D$ with $x = c+d$ and since C and D are cuts, they have no greatest elements. Hence there is $c' \in C$ with $c < c'$ and there is $d' \in D$ with $d < d'$. Thus $x < c' + d'$ so that x is not the greatest element of $C+D$.

Exercises from Chapter 4

4.6 If $X \subset Y$, then the inclusion map from X to Y is one-to-one so that $|X| \leq |Y|$. If, moreover, there is a one-to-one map $f: Y \to X$ then $|Y| \leq |X|$. In view of Theorem 4.5, we conclude that $|X| = |Y|$.

4.14 The map g is one-to-one. Assume $g(n) = g(p)$. If $n \neq 1$ and $p \neq 1$, then $f(n-1) = f(p-1)$ and f is one-to-one, so $n-1 = p-1$ and thus $n = p$. If either n or p is 1, then $g(n) = g(p) = a$ and thus $n = p = 1$ by definition of g. The map g is onto. Indeed, $a = g(1) \in g[\mathbb{N}]$ and if $x \in X$, there is $n \in \mathbb{N}$ with $f(n) = x$ because f is onto. Hence, $x = g(n+1)$ and g is onto.

The map h is one-to-one. Assume $h(n) = h(p)$. If n and p are both even, then $f(\frac{n}{2}) = f(\frac{p}{2})$ and f is one-to-one so $\frac{n}{2} = \frac{p}{2}$ and we conclude that $n = p$. If n and p are both odd, then $g\left(\frac{n+1}{2}\right) = g\left(\frac{p+1}{2}\right)$ so that $\frac{n+1}{2} = \frac{p+1}{2}$ because g is one-to-one. Thus $n = p$. Finally, note that n and p cannot have different parity for otherwise, one of $h(n)$ and $h(p)$ belongs to X and the other to Y, and X and Y are disjoint so that $h(n) \neq h(p)$. Moreover, h is onto, for if $x \in X$ there is $n \in \mathbb{N}$ with $f(n) = x$ and thus $x = h(2n)$, and if $y \in Y$ there is $n \in \mathbb{N}$ with $g(n) = y$ and then $y = h(2n-1)$.

4.18 The map ℓ is onto: for every $(x,y) \in X \times Y$, there is $n \in \mathbb{N}$ and $p \in \mathbb{N}$ with $g(n) = x$ and $h(p) = y$, because g and h are both onto. Thus $\ell(n,p) = (x,y)$. The map ℓ is also one-to-one. Indeed, if $\ell(n,p) = \ell(q,r)$, then $g(n) = g(q)$ and $h(p) = h(r)$. Because g and h are both one-to-one, we conclude that $n = q$ and $p = r$, that is, $(n,p) = (q,r)$.

4.21 A polynomial of degree n in standard form depends on $(n+1)$ coefficients, here chosen in $\mathbb{Z}$. Hence the map $f: \mathbb{Z}^{n+1} \to P_n$ defined by $f(a_0, a_1, \ldots a_n) = p$ where $p(x) = a_n x^n + a_{n-1} x^{n-1} + \ldots a_1 x + a_0$ is onto and thus $|P_n| \leq |\mathbb{Z}^{n+1}|$. In view of Corollary 4.15 and Theorem 4.17, we have $|\mathbb{Z}^{n+1}| = \aleph_0$ by an immediate induction ($\mathbb{Z} = \mathbb{Z}^1$ is countable by Corollary 4.15, and if $\mathbb{Z}^n$ is countable, then $\mathbb{Z}^{n+1} = (\mathbb{Z}^n) \times \mathbb{Z}$ is countable too, by Theorem 4.17). Thus, $|P_n| \leq \aleph_0$, that is, P_n is countable. Moreover, each polynomial has a degree, so that $P_{\mathbb{Z}} = \bigcup_{n \in \mathbb{N}} P_n$ is a union of countably many countable sets, hence is countable by Corollary 4.20.

4.28 Lemma 4.27 shows that if f is one-to-one, then so is g. If moreover f is onto and $C \subset Y$, then

$$C = f[f^{-1}[C]]$$

and thus $C = g(f^{-1}[C])$, that is, g is also onto.

Exercises from the Appendix

A.4 Let $E = \{2n : n \in \mathbb{N}\} \cap \{1, 2, 3, \ldots 100\}$, $T = \{3n : n \in \mathbb{N}\} \cap \{1, 2, 3, \ldots 100\}$ and $F = \{5n : n \in \mathbb{N}\} \cap \{1, 2, \ldots 100\}$. Then $|E| = 50$, $|T| = 33$, and $|F| = 20$. We are looking for the cardinality of $\{1, \ldots 100\} \setminus (E \cup T \cup F)$. Moreover,

$$\begin{aligned}
|E \cap T| &= |\{6n : n \in \mathbb{N}\} \cap \{1, \ldots 100\}| = 16 \\
|E \cap F| &= |\{10n : n \in \mathbb{N}\} \cap \{1, \ldots 100\}| = 10 \\
|T \cap F| &= |\{15n : n \in \mathbb{N}\} \cap \{1, \ldots 100\}| = 6 \\
|T \cap F \cap E| &= |\{30n : n \in \mathbb{N}\} \cap \{1, \ldots 100\}| = 3
\end{aligned}$$

According to Theorem A.1,

$$|\{1, \ldots 100\} \setminus (E \cup T \cup F)| = 100 - (50 + 33 + 20) + (16 + 10 + 6) - 3 = 26.$$

A.5 Let S be the set of students in the class, M the set of those who like mathematics, E the set of those who like English, and P the set of those who like PE. We have

$$\begin{gathered}
|M| = 18, |E| = 16, |P| = 12 \\
|M \cap E| = 7, |M \cap P| = 5, |E \cap P| = 3 \\
|M \cap E \cap P| = 2.
\end{gathered}$$

According to Theorem A.1, the number of students who do not like any of these 3 subjects is

$$\begin{aligned}
|S \setminus (M \cup E \cup P)| &= |S| - (|M| + |E| + |P|) + (|M \cap E| + |M \cap P| + |E \cap P|) \\
&\quad - |M \cap E \cap P| \\
&= 40 - (18 + 16 + 12) + (7 + 5 + 3) - 2 \\
&= 7.
\end{aligned}$$

References

[1] Jean-Pierre Aubin and Hélène Frankowska, *Set-Valued Analysis*, Springer Science & Business Media, 2009.

[2] Guram Bezhanishvili and Wesley Fussner, *An Introduction to Symbolic Logic*, Convergence (2013), DOI:10.4169/loci003990, https://www.cs.nmsu.edu/historical-projects/Projects/symbolic_logic8.pdf.

[3] Guram Bezhanishvili and Eachan Landreth, *An Introduction to Elementary Set Theory*, Convergence (2013), DOI:10.4169/loci003991, https://www.cs.nmsu.edu/historical-projects/Projects/20920110331SetTheoryRevised.pdf.

[4] Garrett Birkhoff, *Lattice Theory*, American Mathematical Society Colloquium Publications, vol. 25, 1967.

[5] John Boyd-Brent, *Harmony and Proportion*, About Scotland Arts pages, http://www.aboutscotland.co.uk/harmony/prop.html.

[6] Chris Caldwell, *Euclid's proof of the infinitude of primes*, https://primes.utm.edu/notes/proofs/infinite/euclids.html.

[7] Peter J. Cameron, *Sets, Logic and Categories*, Springer Science & Business Media, 2012.

[8] Krzysztof Ciesielski, *Set Theory for the Working Mathematician*, London Mathematical Society Student Texts, Cambridge University Press, 1997.

[9] Paul J. Cohen, *Set Theory and the Continuum Hypothesis*, Courier Corporation, 2008.

[10] John Corcoran, *Cambridge dictionary of philosophy "universe of discourse"*, Cambridge University Press, 1995.

[11] Szymon Dolecki, *Analyse Fondamentale: Espaces Métriques, Topologiques et Norrmés*, 2nd ed ed., Hermann, 2013.

[12] Szymon Dolecki and Frédéric Mynard, *Convergence Foundations of Topology*, World Scientific, 2016.

[13] Bob A. Dumas and John E. McCarthy, *Transition to Higher Mathematics: Structure and Proof*, 2nd ed., http://openscholarship.wustl.edu/books/10/, Washington University in St. Louis, 2015.

[14] Gerhard Gierz, Karl Heinrich Hofmann, Klaus Keimel, Jimmie Lawson, Michael Mislove, and Dana Scott, *Continuous Lattices and Domains*, Encyclopedia of Mathematics, vol. 93, Cambridge University Press, 2003.

[15] Leonard Gillman, *Two classical surprises concerning the axiom of choice and the continuum hypothesis*, Amer. Math. Monthly **109** (2002), 544–553.

[16] George Grätzer, *General Lattice Theory*, Springer Science & Business Media, 2002.

[17] George Grätzer, *Lattice Theory: First Concepts and Distributive Lattices*, Courier Corporation, 2009.

[18] George Grätzer, *Lattice Theory: Foundation*, Springer Science & Business Media, 2011.

F. Mynard, *An Introduction to the Language of Mathematics*,
https://doi.org/10.1007/978-3-030-00641-9

[19] Richard H. Hammack, *Book of Proof*, http://www.people.vcu.edu/~rhammack/BookOfProof/, Richard Hammack, 2013.

[20] Seymour Lipschutz, *Set Theory and Related Topics*, 2nd ed., Schaum's Outline, McGraw Hill, 1998.

[21] Làszlò Lovász, Jòzsef Pelikán, and Katalin Vesztergombi, *Discrete Mathematics, Elementary and Beyond*, Undergraduate texts in mathematics, Springer, 2003.

[22] Yiannis Moschovakis, *Notes on Set Theory*, 2nd ed., Undergraduate texts in mathematics, Springer, 2006.

[23] Frédéric Mynard, *Fred Mynard's Webpages-Lyx*, https://sites.google.com/site/fredmynardswebpage/lyx, (2018).

[24] Jorge Picado and Aleš Pultr, *Frames and Locales Topology without points*, Frontiers in Mathematics, Birkhäuser/Springer Basel AG, Basel, 2012.

[25] Julie Rowlett, *Blast into Math*, 1st ed., http://bookboon.com/en/blast-into-math-ebook, Bookboon.com, 2013.

[26] Bertrand Russell, *The Problems of Philosophy*, Oxford paperback university series, no. Opus 18, Oxford University Press, 1959.

[27] Douglas Smith, Maurice Eggen, and Richard St Andre, *A Transition to Advanced Mathematics*, Nelson Education, 2014.

[28] Daniel Solow, *How to Read and Do Proofs: An Introduction to Mathematical Thought Processes*, 6th ed., Wiley, 2013.

List of Symbols

2^κ	cardinality of the power set of a set of cardinality κ
$<$	$a < b$ if $a \leq b$ and $a \neq b$ in a poset
$\#$	meshing relation on $\mathbb{PP}X$
$\aleph_0$	cardinality of $\mathbb{N}$
$\aleph_1$	smallest cardinality greater than $\aleph_0$
$\bigcap$	intersection (infinite)
$\bigcup$	union (infinite)
$\bigvee A$	supremum (least upper bound) of A
$\bigwedge A$	infimum (greatest lower bound) of A
$\binom{n}{k}$	n choose k: number of subsets with k elements of a set with n elements
$\perp$	least element of a complete lattice
$\cap$	intersection (finite)
$\cdot^c$	set-theoretic complement
χ_A	indicator function of A
$\cup$	union (finite)
$\emptyset$	empty set
$\equiv$	logical equivalence
$\equiv_p$	congruence of integers modulo p
$\exists$	existential quantifier 'there is'
$\forall$	universal quantifier 'for all'
$\gcd(a,b)$	greatest common divisor of a and b
$\iff$	biconditional statement: "if and only if"
$\in$	membership relation: $x \in A$ if x is a member of the set A
$\inf A$	infimum (greatest lower bound) of A
$\lceil \cdot \rceil$	ceiling function
$\lfloor \cdot \rfloor$	floor function
$\Longrightarrow$	conditional statement: "implies"
$\mapsto$	maps to

F. Mynard, *An Introduction to the Language of Mathematics*,
https://doi.org/10.1007/978-3-030-00641-9

$\mathbb{N}$	set of natural numbers
$\mathbb{P}X$	powerset of X
$\mathbb{Q}$	set of rational numbers
$\mathbb{R}$	set of real numbers
$\mathbb{Z}$	set of integers
$\neg$	negation
$\not\mid$	does not divide
$\notin$	non-membership relation: $x \notin A$ if x is not a member of the set A
$\oplus$	exclusive or (XOR)
$\setminus$	set-theoretic difference
$\sim_{\mathscr{P}}$	equivalence relation associated with a partition $\mathscr{P}$
$\sim_f$	equivalence relation on the domain of an onto map f associated with f
$\subset$	subset relation
$\sup A$	supremum (least upper bound) of A
$\tilde{f}$	function f seen as a relation (graph)
$\tilde{f}$	graph of f
$\top$	greatest element of a complete lattice
$\triangle$	symmetric difference (between sets)
$\vee$	logical disjunction
$\lvert X \rvert$	cardinality of X
$\wedge$	logical conjunction
$\{\ \}$	set builders
$\{x \in X : \varphi(x)\}$	set of elements of X such that $\varphi(x)$
$a \mid b$	a divides b
C_q	Dedekind cut associated with q
$f[A]$	image under f of the subset A of the domain of f
$f^{-1} : Y \to X$	inverse function of a bijective function $f : X \to Y$
$f^{-1}[B]$	preimage under f of the subset B of the codomain of f
$f^{-1}[y]$	fiber of y under f: $\{x \in X : f(x) = y\}$
I_X	identity relation of X
i_X	identity function of X
$R(A)$	image of $A \subset X$ under the relation $R \subset X \times Y$
$R(x)$	image of x under the relation R
$R^*(A)$	polar of A with respect to R
R^{-1}	inverse relation of R
$S \circ R$	composite of the relations R and S
$W_{\downarrow x}$	initial segment of the well-ordered set W corresponding to x
X/R	quotient set of X under the equivalence relation R
$X \simeq Y$	X is order isomorphic to Y
$X \times Y$	Cartesian product of X and Y
Y^X	set of functions from X to Y

Index

F. Mynard, *An Introduction to the Language of Mathematics*,
https://doi.org/10.1007/978-3-030-00641-9